Blueprint Reading for
COMMERCIAL CONSTRUCTION

Charles D. Willis

Blueprint Reading for
COMMERCIAL CONSTRUCTION

Charles D. Willis

10 9 8 7 6 5 4 3

LIBRARY OF CONGRESS CATALOG CARD NUMBER: 77-87887
ISBN: 0-8273-1654-2
Printed in the United States of America.
Published simultaneously in Canada by
Delmar Publishers, a Division of
Van Nostrand Reinhold, Ltd.

DELMAR PUBLISHERS INC. • ALBANY, NEW YORK 12205

Preface

Construction is the largest industry in the United States. Its labor force exceeds fifteen percent of the nation's total working force. Blueprint reading is of course essential for anyone working in construction. This text offers not only the basics of blueprint reading, but it also includes the fundamentals of construction methods that make blueprint reading meaningful.

The text is divided into nine sections with a total of thirty-two units of instruction. The first ten units serve as introduction and present the basics of blueprint reading. The remaining units are specialized according to types of work.

Each unit is preceded by objectives which state what the student is responsible for in that unit. The unit material focuses on material which aids the student in fulfilling these objectives. The unit information is reinforced by review questions many of which engage the student in actively reading and sketching blueprint drawings and plans. Having completed the units, the student should have a firm foundation for reading commercial blueprints.

Technical terms are carefully defined in each unit as they appear. In addition, the text includes a glossary which serves as a ready reference.

The author of this textbook, Charles D. Willis, has taught construction classes at John Brown University for twenty-five years. His classes have included Architectural Drafting and Design; Structural Design in Wood, Steel, and Reinforced Concrete; Estimating; Construction Management; Vocational Shop; and Cabinetmaking. He is currently the head of the Department of Building Construction at that institution. He has designed and supervised the construction of several commercial construction projects. He is a member of the American Institute of Constructors, the National Fire Protection Association, and the Building Research Institute.

A current catalog including prices of all Delmar educational
publications is available upon request. Please write to:

Catalog Department
Delmar Publishers Inc.
50 Wolf Road
Albany, New York 10025

CONTENTS

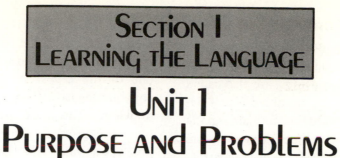

SECTION I
LEARNING THE LANGUAGE

UNIT 1
PURPOSE AND PROBLEMS

OBJECTIVES

After studying this unit, the student should be able to:

- tell why it is important to study blueprint reading.
- explain the role of the architect and engineer.
- identify plans and specifications.
- explain the importance of symbols used in blueprints.

CONSTRUCTION —
OUR LARGEST INDUSTRY

Over the years, construction has steadily become this country's largest industry. This is true when all who work directly or indirectly are included. The number now employed in construction exceeds 15 percent of the total working force of the nation. New construction accounts for about two thirds of the annual volume with repairing, remodeling, and maintenance making up the balance. The workers who manufacture and distribute building materials and equipment are examples of indirect construction workers. In a sense, everyone is affected by this great industry because people must have buildings in which to live, work, and worship. They must have highways, power lines, and utilities to provide transportation and energy.

The classifications of construction most generally understood include buildings, highways, and heavy construction. Blueprint reading is an essential skill for anyone working in construction. This book is written especially for people who specialize in the construction of buildings, but the principles apply to highway and heavy construction as well. A special effort is made to include topics representing many construction methods not found in single-family residences but important in larger buildings. For this reason the term "Commercial" is included in the title.

This material is applicable to the carpenter, the mason, the plumber, the drafter, the owner, the estimator, the building inspector, the electrician, the contractor, etc. Anyone interested in reading blueprints for construction can profit by a careful study of the principles included in this book.

ROLE OF THE ARCHITECT
AND ENGINEER

Architects are designers who prepare the drawings for building many types of projects. They work for owners who want buildings designed and built to meet specific needs. Because each building is unique, that is, one of a kind, complete sets of directions must be made for each building. At times similar buildings are designed but built on different lots, so they still need special plans. Architects often hire engineers to help prepare some parts of the instructions. Typical parts designed by engineers are structure, and mechanical and electrical systems. The architect supervises all phases and is responsible to the owner for the design.

PLANS AND SPECIFICATIONS

The instructions for building a building are of two types, written and graphic, figure 1-1. The graphics, or drawings, are also called plans, working drawings, or blueprints. Because of their great value in explaining the physical nature of an object, drawings are used extensively. However, there are instructions that are better written than drawn. These are prepared in the form of a book or manual called the *specifications* or just "specs." Usually the term "specs" will be used in this book in place of specifications since it is a term much used in the field. Specs add to the information shown on the drawings and are every bit as important to the builder. Units 7, 8, and 9 give more details about the plans and specs.

BLUEPRINT READING

The process of interpreting the architect's instructions may be defined as *blueprint reading.* This process is called blueprint reading regardless of the type of drawing reproduction method used. The term blueprint comes from one of the oldest methods, one in which a copy in negative form is made of the original drawing. The lines appear white on a dark blue background, figure 1-2. This process involves a water rinse and drying cycle. This wet process is not used very much now because it requires too many steps.

A dry process called *Diazo* makes copies called blue line prints or white prints. These are popular because they are easy to make and show the drawings with blue lines on a white background. Another type of paper used shows black lines on a white background, figure 1-3, page 4. Ammonia fumes, used for years to develop these Diazo prints, require a warm up time and their odor is offensive. An odorless pressure Diazo copying process which is now in use makes the Diazo white prints even more popular, figure 1-4, page 5.

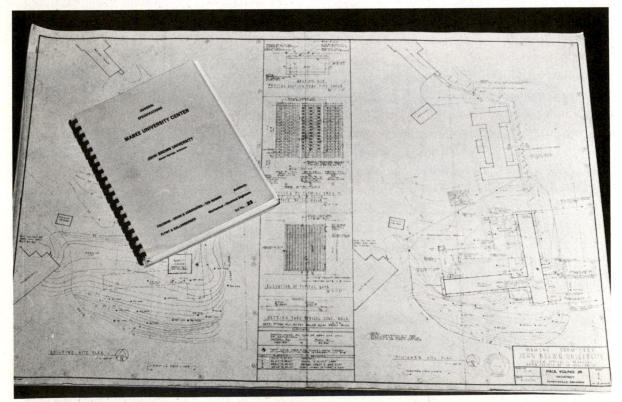

Fig. 1-1 Specifications and plans

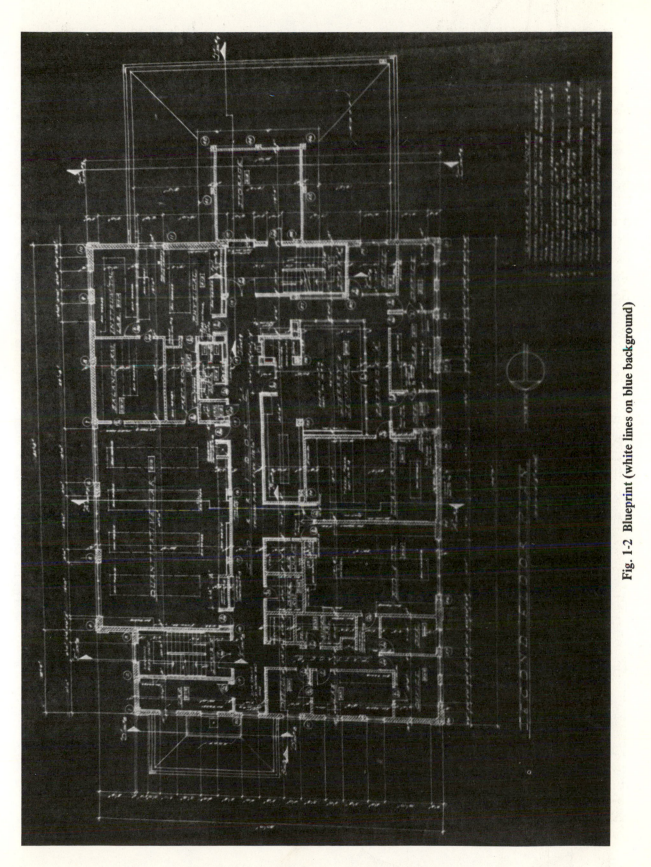

Fig. 1-2 Blueprint (white lines on blue background)

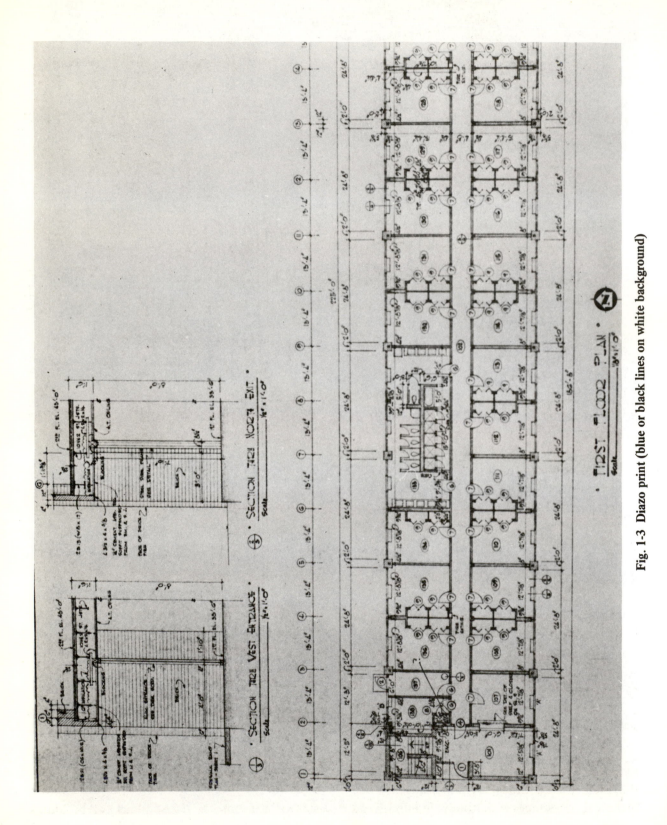

Fig. 1-3 Diazo print (blue or black lines on white background)

Fig. 1-4 Pressure Diazo copier *(Bruning Division, Addressograph Multigraph, Schaumburg, IL)*

Since the specs supplement the drawings, it is necessary to include them in this text. The construction person who wants to advance to supervisor or higher in the field must learn to interpret the specs as well as the drawings.

A UNIVERSAL LANGUAGE MODIFIED

Although pictures and drawings are the same for a viewer from any nation, the message they tell is sometimes vague. *Blueprints* are technical drawings with many symbols and conventions. *Symbols* are figures or signs that stand for some part of the building. *Conventions* are standardized ways of doing things, like the ways doors and windows are scheduled in the plans.

Blueprints for buildings use symbols to a greater degree than blueprints for machines and other products. There are two good reasons for this. The first reason has to do with the size of the drawing compared to the size of the object. Most plans for buildings are so small it is impossible to show all the parts. Hence, symbols are used to represent certain parts. The second reason for more symbols is the need for speed and economy in drawing the thousands of parts in the building. The drawings are used only once to build one building. It would be a waste of time to make exact drawings of every part, so symbols are used instead.

DEFINING THE PROBLEM

The problem then is to read or interpret the instructions prepared by the architect in order to build the building the owner wants. The builder is expected to bring the job expertise and trade skills consistant with the scope of the project. There are many instances in which the builder has to figure a way to perform a task. The architect often is concerned more with the end result than with the step-by-step process. Blueprint reading is an important step toward the successful communication between the designer and the builder. Working together they can meet the owner's need for a new structure.

SUMMARY

- Architects and engineers are professional designers who prepare plans and specifications for construction.

- Both graphic and written instructions are needed to describe completely a project.

- Blueprint reading is the process of interpreting the plans and specifications.

- Blueprints are not all blue; white prints with blue or black lines are now more common.

- Building blueprints are technical drawings which use many symbols to convey construction information.

- The builder is expected to follow the instructions of the architect and to contribute skills in a joint effort to build new structures.

REVIEW QUESTIONS

1. What is our nation's largest industry?

2. Why should the construction person study blueprint reading?

3. Who prepares the drawings and specifications for building projects?

4. Engineers prepare instructions for parts of a building project. Name two parts.

5. Give two other names for drawings.

6. What is a simpler term for specifications?

7. What is the term given to the process of interpreting the architect's instructions?

8. What drawing reproduction process makes white lines on a blue background?

9. The Diazo process makes prints with a white background. What color are the lines?

10. Why do blueprints for buildings use symbols more than do blueprints for machines?

11. How many buildings usually are built from a set of blueprints?

12. What does the architect expect the builder to contribute to the project?

Unit 2
Alphabets and Symbols

OBJECTIVES

After studying this unit, the student should be able to:

- identify the catalog of lines used in drawings.
- give different uses for the lines in drawings.
- tell what drawing techniques make architectural drawings more artistic.
- sketch symbols commonly used in building drawings.

ALPHABETS

To read blueprints it is necessary first to learn the language in which they are written. Each written language has its alphabet from which words are formed. Our twenty-six letter English alphabet comes down to us from the Greek alphabet as perfected by the Romans. But the earliest forms of written language often have picture symbols for words and ideas. Two such examples are the Egyptian hieroglyphics and the Chinese language, figures 2-1, 2-2.

Although pictures are a universal language understood by all, technical drawings have

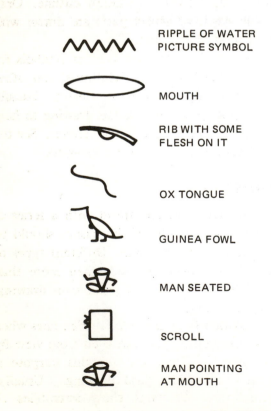

RIPPLE OF WATER
PICTURE SYMBOL

MOUTH

RIB WITH SOME
FLESH ON IT

OX TONGUE

GUINEA FOWL

MAN SEATED

SCROLL

MAN POINTING
AT MOUTH

(SCROLL)

DOCUMENT

(MAN)

SCRIBE

THE MEANING OF THIS
WORD DEPENDS UPON
THE LAST SYMBOL —
(SCROLL OR MAN)

(MAN)

TO PRAY
MAN POINTING TO MOUTH

Fig. 2-1 Egyptian hieroglyphics

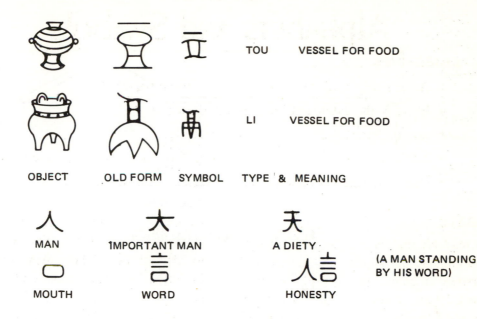

OBJECT   OLD FORM   SYMBOL   TYPE & MEANING

TOU   VESSEL FOR FOOD

LI   VESSEL FOR FOOD

MAN   IMPORTANT MAN   A DIETY   (A MAN STANDING
BY HIS WORD)

MOUTH   WORD   HONESTY

Fig. 2-2 Chinese letters

special features that must be studied. The most important feature of technical drawings is the line. The types of lines used are so varied a catalog of them is called an *alphabet of lines.*

LINE TYPES IN THE ALPHABET OF LINES

Object Lines

The basic line used to show the outlines and edges of objects is a continuous solid line. The weight of the line, whether it is heavy, medium, or light, is an important variable. The size of the drawing is a factor, but the complexity of the object determines the line weight to use. A simple object is drawn with one weight of line, medium or heavy weight, as in figure 2-3, page 10. When more detail is included, lighter-weight lines are used in the order of the importance of the element. Note in figure 2-4, page 10, that a window is added to the building shape of figure 2-3. The window, being less important than the outline of the house, is drawn with slightly thinner lines.

The window parts should be drawn with thinner lines than the window outline. Only visible edges of object parts are drawn with these solid object lines.

Solid lines are also used as symbols for certain types of piping, contours, and other parts not really seen in the object. Usually there will be a note on the drawing to help the reader understand the difference. See the architectural symbols in the appendix.

Dotted Lines

Dotted lines are drawn with a series of dashes. The length of the dashes should be the same in any one line. Different types of dotted lines may be used when more than one element is indicated in the same drawing, figure 2-5, page 10.

Dotted lines are called hidden lines when they represent edges concealed from view by the object. Their use for this purpose is limited in architectural drawing. Usually, architectural drawings show assemblies of many parts. When dotted lines are used to

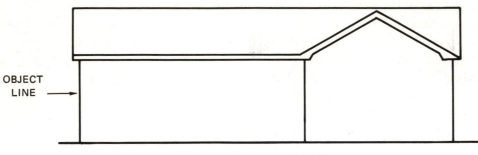

FRONT VIEW OF BUILDING

Fig. 2-3 One weight of line drawing

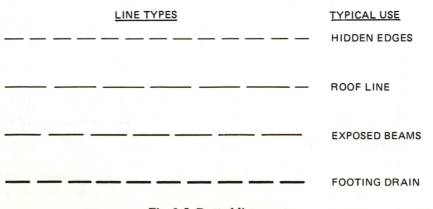

HOUSE WITH WINDOWS

Fig. 2-4 Drawing with line variations

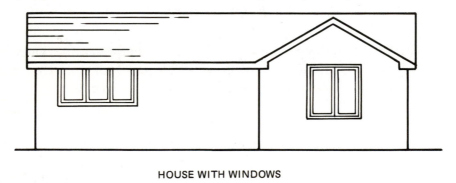

LINE TYPES	TYPICAL USE
	HIDDEN EDGES
	ROOF LINE
	EXPOSED BEAMS
	FOOTING DRAIN

Fig. 2-5 Dotted lines

reveal all concealed edges, the drawing becomes very hard to read. An example of the use of hidden lines in building plans is a footing outline as shown in figure 2-6. The footing outline cannot be seen in the top views because the footings are underground. They are shown using dotted hidden lines.

The dotted line also is used to show the location of edges that occur above the plane of the drawing. For example, the overhang of a roof is shown with a dotted line on a floor plan drawing. When the viewer looks at the floor of a building, the roof is above the field of vision. The outline of the roof is important information shown on the drawing representing the floor, figure 2-7.

Exposed beams, changes in ceiling height, and other important ceiling parts are shown as dotted lines on the floor plan. Also, some wiring symbols, welded wire mesh, footing drains, and other items are drawn using dotted lines. Most dotted lines are medium to light in weight. Footing drains can be drawn with heavyweight lines.

Center Lines

Center lines are drawn as long and short dashes, figure 2-8, page 12. They are used to indicate the center of a round element to show that an object is symmetrical about it, figure 2-9, page 12. Center lines are used

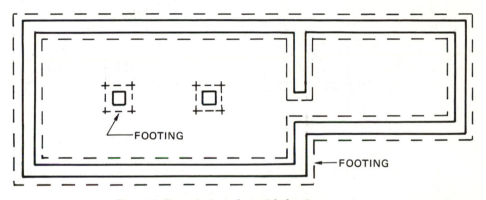

Fig. 2-6 Foundation plan with footings

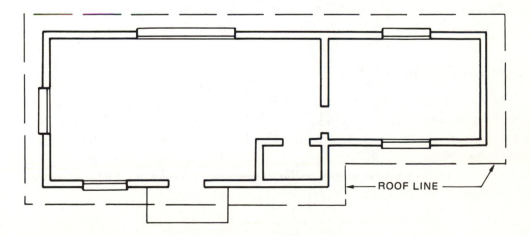

Fig. 2-7 Floor plan with roof line

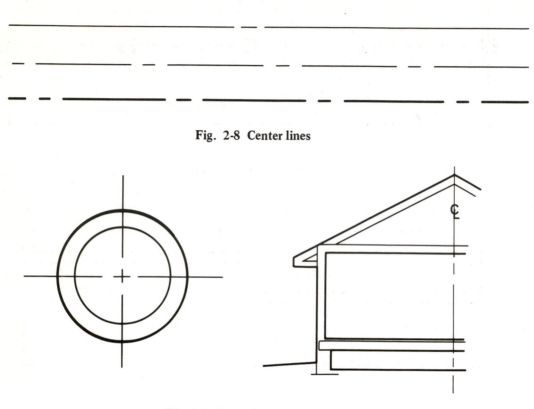

Fig. 2-8 Center lines

Fig. 2-9 Center lines on objects

to indicate the centers of structural parts such as columns, figure 2-10. Another use of the center line is to indicate steel joists on framing plans.

Extension and Dimension Lines

Extension lines are light lines which project from an object to a convenient place for dimensioning, figure 2-11. They should not touch the object.

Dimension lines are light unbroken lines drawn between extension lines and labeled with the distance between them, figure 2-12. Dimension lines may also run to center lines, dotted lines, and object lines.

At the end of the dimension line, arrowheads are required to indicate exactly what is being dimensioned. Several forms of arrowheads are used in architectural drawings, figure 2-13, page 14. The choice of arrowheads used belongs to the person working on the drawings except on drawings dimensioned by modular drafting rules. (Unit 6 has more information about modular drafting conventions.)

The extension lines should extend beyond the dimension line so they do not look like an object corner. Dimension lines terminated with dots or slashes should also extend slightly past the extension line. Conventional arrowheads stop with their points just touching the extension lines, figure 2-13, page 14.

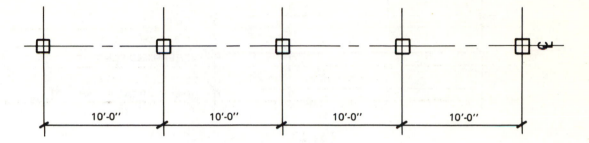

Fig. 2-10 Columns on center line

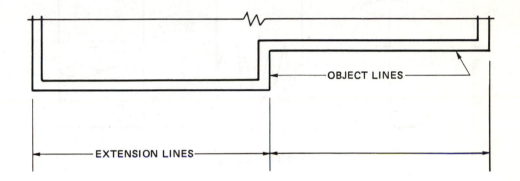

Fig. 2-11 Extension lines

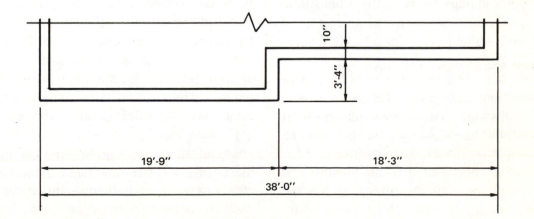

Fig. 2-12 Dimension lines

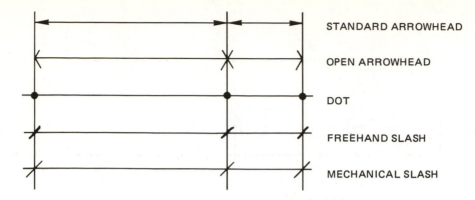

Fig. 2-13 Types of arrowheads

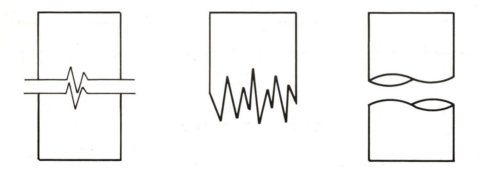

Fig. 2-14 Break lines

OTHER LINES

Break lines are used to indicate that something has been left out. They may end a detail when only a partial view is desired. They are found on detail drawings where there is not enough room to draw the whole detail. Another common example is their use to end a flight of stairs on a floor plan, figure 2-14.

The same convention is used in dimension lines that cross a cut plane where something is missing. The correct dimension is indicated, but the dimension line is shortened to match the object as cut, figure 2-15.

Another line related to the break line is the match line. Its function is much the same except it is used to stop a drawing too large for one sheet. A similar match line is drawn on the next sheet to continue the drawing; therefore, nothing is left out. If the two sheets are joined at the match line, they form a complete drawing. The match line is drawn as a continuous line or as a center line. No zigzag form is used as in the break line, figure 2-16.

Cutting plane lines are used to indicate an imaginary plane cut through the object, figure 2-17. The construction of the object is thus revealed in a section view of that plane. The lines vary in form but always have some identification at their ends, figure 2-18, page 16.

Guidelines are lightweight lines used to locate freehand lettering, figure 2-19, page 16.

Leaders are also lightweight straight lines used to connect notes to the parts of the object. Leaders terminate in an arrowhead, figure 2-19.

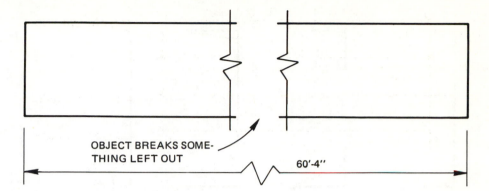

Fig. 2-15 Broken dimension line

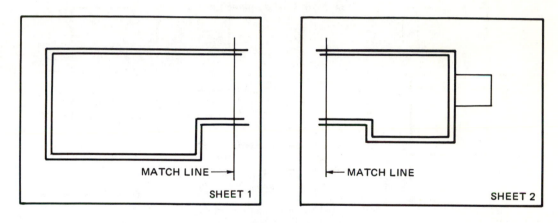

Fig. 2-16 Match line example

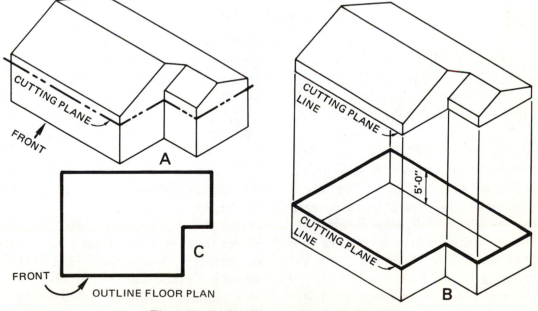

Fig. 2-17 Cutting plane passing through building

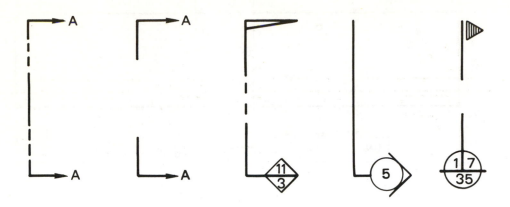

Fig. 2-18 Cutting plane lines

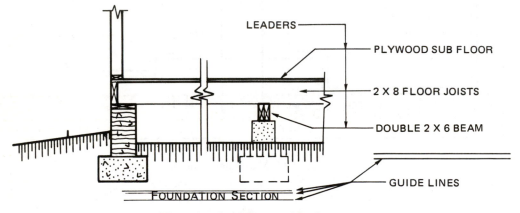

LEADERS

PLYWOOD SUB FLOOR

2 X 8 FLOOR JOISTS

DOUBLE 2 X 6 BEAM

GUIDE LINES

FOUNDATION SECTION

Fig. 2-19 Guidelines and leaders

In addition to the lines drawn with drafting instruments, architectural drawings may have freehand lines for some details. Among freehand details are shrub and tree symbols, contours, special materials, etc. These details, along with the many line weights used for object lines, tend to make drawings much more artistic. The slight crossing of object lines at external corners adds to the effect. Freehand lettering, if well done, also enhances a drawing and is typical in blueprints for buildings.

SYMBOL TYPES

In addition to an alphabet of lines, architectural drawings have a number of symbols which look more or less like the objects they stand for. Plumbing fixtures are shown by symbols of this type, figure 2-20. Other symbols are used to show the kind of material used. They usually have no relationship to the shape or kind of material used. When used consistently, they give important facts about the nature of the object, figure 2-21. A third type of symbol has an assigned meaning somewhat related to the way the object looks. Examples of this type symbol are electrical outlets shown in figure 2-22.

Symbols used in architectural blueprints are given in the appendix of this text.

WATER CLOSET
LAVATORY
BATH TUB
WASHER
DRYER

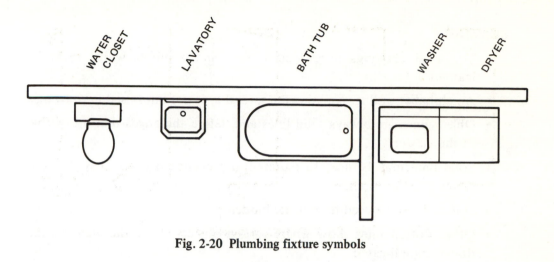

Fig. 2-20 Plumbing fixture symbols

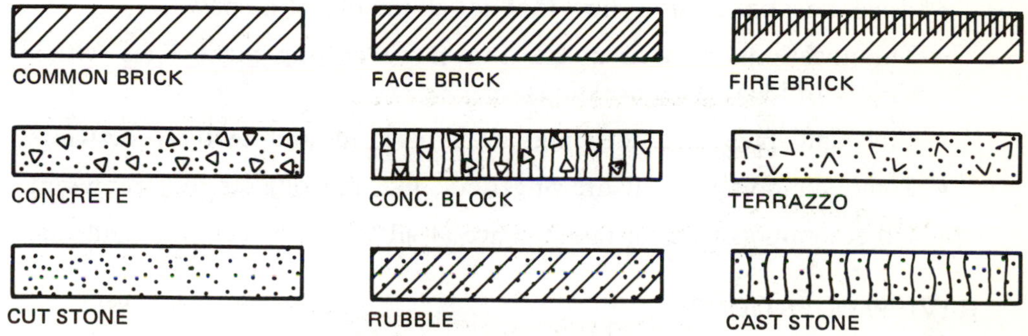

COMMON BRICK
FACE BRICK
FIRE BRICK
CONCRETE
CONC. BLOCK
TERRAZZO
CUT STONE
RUBBLE
CAST STONE

Fig. 2-21 Building materials symbols

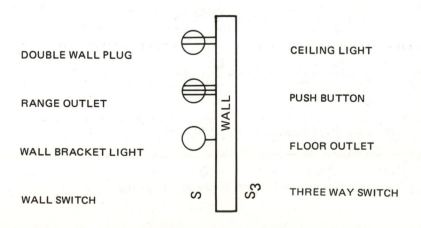

DOUBLE WALL PLUG
RANGE OUTLET
WALL BRACKET LIGHT
WALL SWITCH
WALL
S
S3
CEILING LIGHT
PUSH BUTTON
FLOOR OUTLET
THREE WAY SWITCH

Fig. 2-22 Electrical outlet symbols

The student needs to refer to them many times as blueprint reading skill is developed. After awhile most of them are memorized in the process.

SUMMARY

- Technical drawings have special lines and symbols which must be learned.
- An alphabet of lines contains the several types of lines used in drawings.
- Object lines are always solid lines and define the edges that are visible to the viewer.
- Solid lines may be used to represent objects not visible if they are used as symbols for piping etc.
- Dotted lines are used to indicate hidden edges.
- Other dotted lines show where certain objects above the plane of the drawing are located.
- Several arrowhead types are used on dimension lines.
- Modular drafting rules require designated arrowhead types.
- Break lines are used to show that not all the object is shown.
- The locations of sections are shown by cutting plane lines with a variety of identifying notations.
- Freehand lines and lettering are typical in architectural drawings.
- Some symbols used in drawings look like the objects they represent.
- Other symbols bear no resemblance at all to the objects they indicate.

REVIEW QUESTIONS

1. What is the most important feature of technical drawings?

2. What is the catalog of lines used in drawings called?

3. Which line type is used for outlines and edges of objects?

4. Name one case where solid lines are used as symbols.

5. Which line type is used for hidden edges in a drawing?

6. Give an example of hidden lines used in building plans.

7. Which type line is used to show roof outlines on floor plan drawings?

8. Sketch a center line.

9. Arrowheads are used at the ends of which lines, dimension lines or extension lines?

10. Sketch a break line as used to show that the object is cut off.

11. What weight lines are used for guidelines?

12. What parts of an architectural drawing are drawn using freehand lines?

13. How are architectural drawings made more artistic?

14. What symbols used on drawings look like the objects they represent?

15. Sketch the symbols for brick and concrete.

Unit 3
Basic Views—Single Plane

OBJECTIVES

After studying this unit, the student should be able to:

- sketch multi-view drawings of a simple object.
- identify elevation and plan views of an object.
- recognize the shape of a building from a study of the elevations and plans.

TYPES OF PROJECTION DRAWINGS

Most of the drawings found on blueprints for buildings are developed by the *orthographic projection* method. This method projects the image of the side of an object onto a plane at true size and shape. The projectors of points on the object are perpendicular to the picture plane, figure 3-1.

We normally view objects in *perspective*. A photograph of an object is this type of projection. Perspective drawings are very useful in giving a realistic view of the object but very difficult to measure. In this type drawing, the size and shape of the object are not true. The projectors of points on the object are at various angles to the picture plane converging to a point, figure 3-2.

MULTI-VIEW DRAWINGS

Multi-view drawings show single plane views of the object from different sides.

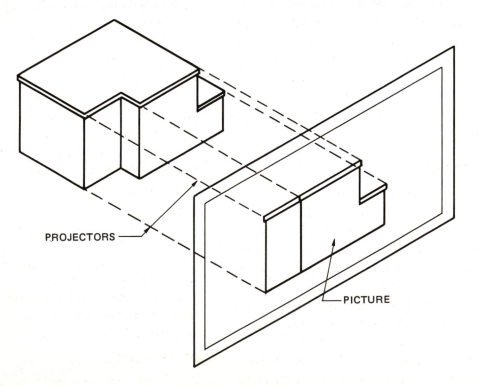

PROJECTORS

PICTURE

Fig. 3-1 Orthographic projection

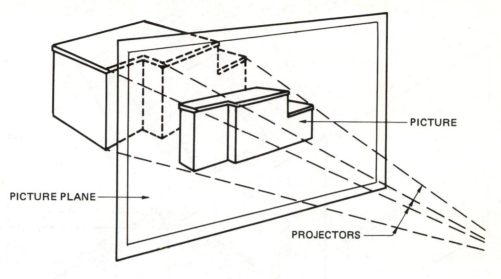

PICTURE

PICTURE PLANE

PROJECTORS

Fig. 3-2 Perspective projectors

Each drawing shows one side as projected by the orthographic method. They are true size and shape and can be measured. They are usually explained by using a folding glass box about the object, figure 3-3, page 22.

The basic views for architectural drawings are the elevations, the top view called the roof plan, and various floor plans. In addition there are sections and details which are presented in Unit 4.

ELEVATIONS

Elevations are side views of the object. They are so named because heights or elevations can be measured on them. For a building there are standard names for the different elevations. When the compass direction the building will face is known, the side of the building facing north is named the north elevation. The other sides are named correspondingly, as in figure 3-4, page 22.

If the compass orientation is not known, then the front side is called the front elevation. The right and left sides are named

similarly using the front or principle elevation of the building as the reference view. The back side is named the rear elevation, figure 3-5, page 23.

It should be noted that sloping surfaces such as the gable roof in figure 3-5, page 23, front elevation, are not true size. This is because they are not parallel to the projection plane on which this elevation is drawn. The true length of the sloping rafter can be measured on the right elevation. The roof area can then be computed accurately by multiplying the rafter length by the length of the roof.

A special drawing called an *auxiliary view* sometimes is used when the true shape of a sloping surface is needed. Auxiliary views are used more in mechanical drawings than in architectural drawings. An auxiliary view of a roof surface is shown in figure 3-6, page 23.

ROOF PLAN

If we view the building from above we see the roof. The orthographic projection

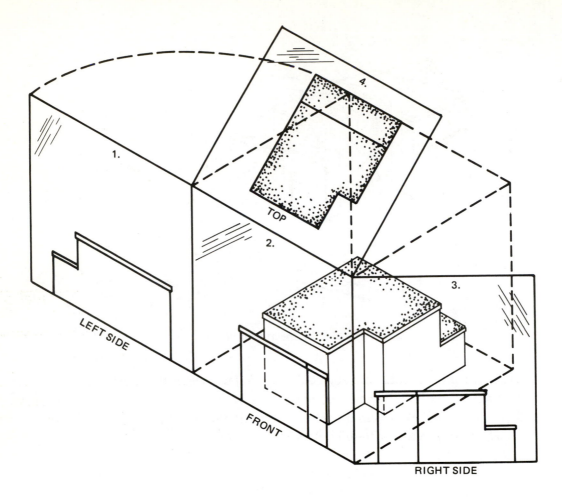

Fig. 3-3 Glass box unfolded

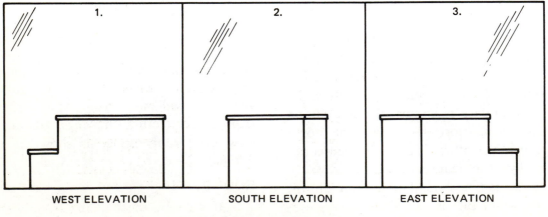

Fig. 3-4 Elevations by compass

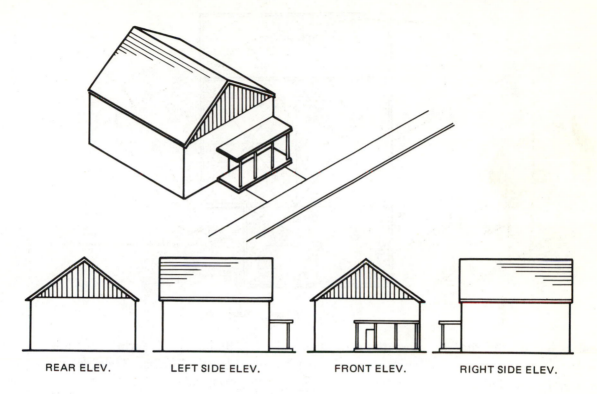

REAR ELEV.   LEFT SIDE ELEV.   FRONT ELEV.   RIGHT SIDE ELEV.

Fig. 3-5 Elevations by side

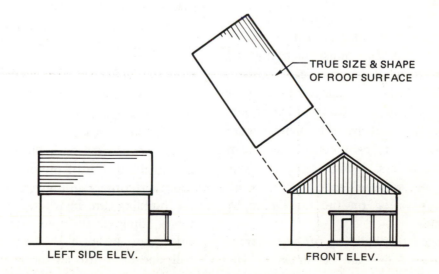

TRUE SIZE & SHAPE
OF ROOF SURFACE

LEFT SIDE ELEV.   FRONT ELEV.

Fig. 3-6 Auxiliary view of roof

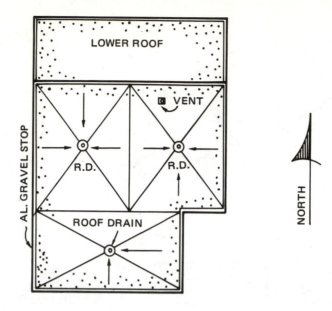

Fig. 3-7 Top view – roof plan

drawing of the roof is called the *roof plan.* Any drawing made looking down on the object is called a *plan view* just as all the side views are called elevations. The roof plan helps define the materials, size, and location of roofing and related parts, figure 3-7.

FLOOR PLANS

Floor plans are drawings made by looking down on the particular floor in question. To do this the drafter must mentally lift off the roof so what is there can be seen. The plane of separation usually is taken four or five feet above the floor. The doors and windows are cut by this plane, figure 3-8, and 3-9. Base cabinets are seen, but upper wall cabinets, being above the separation plane, require special treatment. They are represented on the floor plan by dotted outlines.

By cutting the building off at different levels, various plans can be drawn. The lowest level yields a foundation or basement plan, figure 3-10, page 27.

Building blueprints include plumbing and electrical plans. They show the location of pipes and electrical systems on copies of the basic floor plans, figure 3-11, page 28. They are not true projection drawings because much of the detail shown is concealed in the actual building. When the symbols are understood, they are effective sources of information about these special systems. Units 31 and 32 contain more information about mechanical and electrical work.

PUTTING IT ALL TOGETHER

After assembling the elevations and plans, it is easier to picture the basic form of the building. A simple sketch is another way to get acquainted with the building. (Unit 5 gives a discussion of pictorial views.) Estimators, and others interested in understanding the building, are instructed to walk through the building via the plans. Start with the first floor plan and go in the front door, down the hall, and look into all the rooms. This way a good idea is formed of the arrangement of the whole building in a very short time.

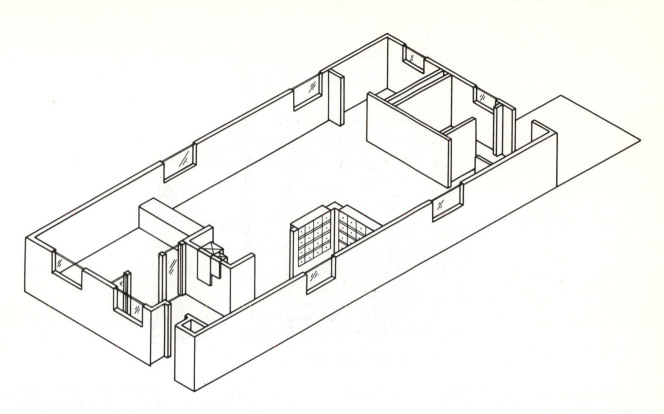

Fig. 3-8 Building cut of a Post Office 4 ft. above floor

SUMMARY

- Drawings for buildings are developed largely by the orthographic method.

- Orthographic views are true in size and shape.

- Perspective views look realistic but cannot be measured easily.

- The basic architectural drawings are the plans and elevations.

- The north elevation of a building is the side of the building facing north.

- Auxiliary views are used to show true shapes of sloping surfaces.

- A floor plan shows the floor as viewed from above with the upper part of the building removed.

- Plumbing and electrical plans show the location of pipes and circuits that are concealed in the construction.

- Visualizing the finished building by studying the plans and elevations is an important step in blueprint reading.

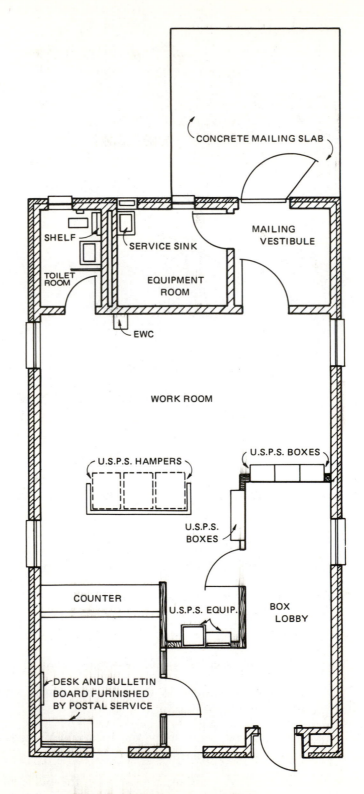

Fig. 3-9 Floor plan of a Post Office

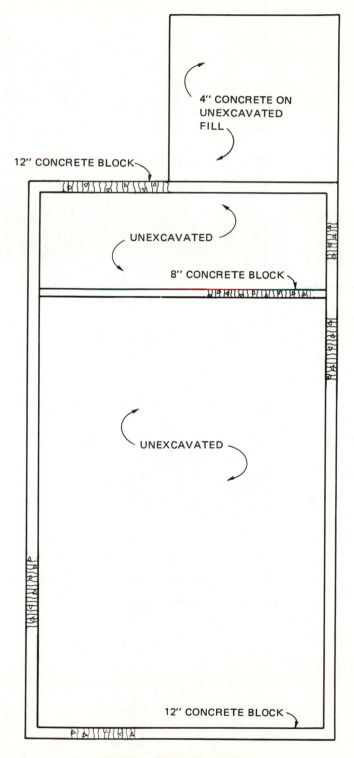

12″ CONCRETE BLOCK

4″ CONCRETE ON UNEXCAVATED FILL

UNEXCAVATED

8″ CONCRETE BLOCK

UNEXCAVATED

12″ CONCRETE BLOCK

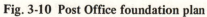

Fig. 3-10 Post Office foundation plan

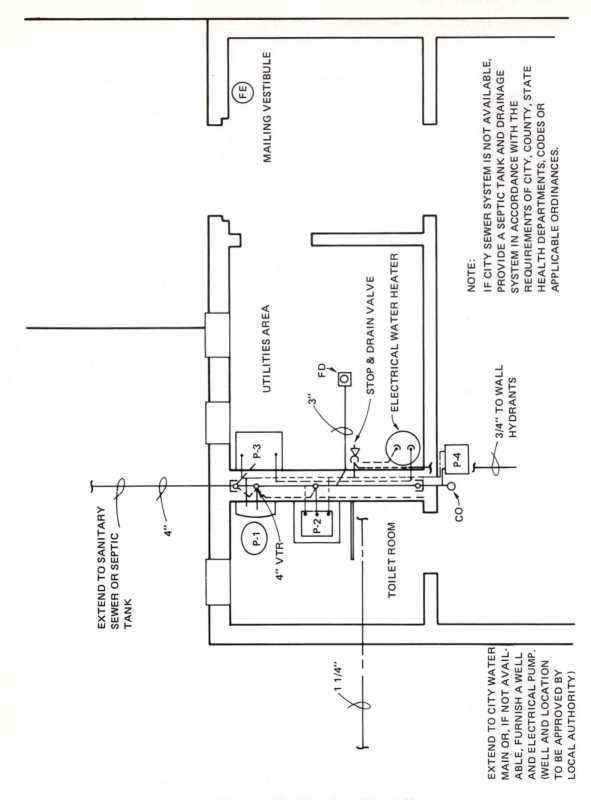

Fig. 3-11 Plumbing plan of Post Office

REVIEW QUESTIONS

1. What are the side views of a building called?

2. If the front of a building faces south, what is the name given to a drawing of the right side?

3. Which views of a building show the interior room arrangements?

4. At what height is the building cut to show the room arrangements?

5. What is the top view of a building called?

6. What do plumbing plans show besides the room arrangements of a building?

7. What is a good way to get acquainted with a new building through the blueprints?

8. Sketch multi-view drawings of these objects.

   8a.

   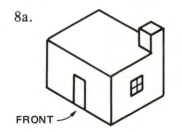

   FRONT

8b.

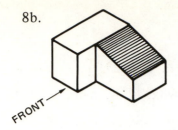

8c.

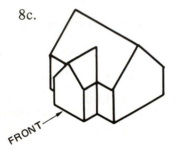

Unit 4
Details and Sections

OBJECTIVES

After studying this unit, the student should be able to:

- tell what is meant by the term details.
- identify section views and their importance in blueprint reading.
- sketch cutting line symbols.
- interpret revolved section views.

DETAILS DEFINED

Details and sections make up a large part of all construction drawings. The term *details* covers the broad category of isolated and enlarged drawings used to explain the construction of the building. Details include enlarged elevations and plans as well as sections where there is a need to explain the object fully. Each detail gives specific information at key points in the building. Sometimes sections may be called details on the plans. This should not be a problem to the blueprint reader since all enlarged drawings of portions of the structure are rightly called details.

SECTIONS DEFINED

Sections are drawings that show the materials and how they fit together at imaginary planes cut through the object. Next to the plan and elevation views, they are probably the most important drawings in blueprints for buildings. In order to construct a building whose appearance is shown by the elevations, it is necessary to know how the pieces go together. The plans and elevations basically show where, while the sections show what and how, the parts are assembled. The materials used are identified in the sections by notes and symbols.

The wall section is a good example of a section that shows the required materials and their relationships, figure 4-1, page 32. This section shows the foundation wall and its footing. It shows the ground line outside the house. The floor framing, wall framing, and roof framing are all indicated. Exterior and interior finish surfaces are shown and noted. Insulation in the walls and in the attic is also noted. Some simple buildings can be detailed adequately with one good wall section.

CUTTING PLANE LINES

Cutting plane line symbols are identified in Unit 2, figure 2-18. Every cutting plane line will have a corresponding section drawing somewhere in the blueprints. Since the sections are so important in reading the blueprint, the architect tries to locate them near their cutting plane line. If possible they are put on the same sheet. Most of the symbols in figure 2-18 are appropriate in this case.

In many cases, however, there is not space for the sections on the sheet where the plane is cut. Also, some sections will apply to more than one sheet, adding to the problem. Special symbols are used for drawings with the complexity just described, figure 4-2, a, page 32. The more elaborate symbol is shown in figure 4-2, b, page 32. It numbers the

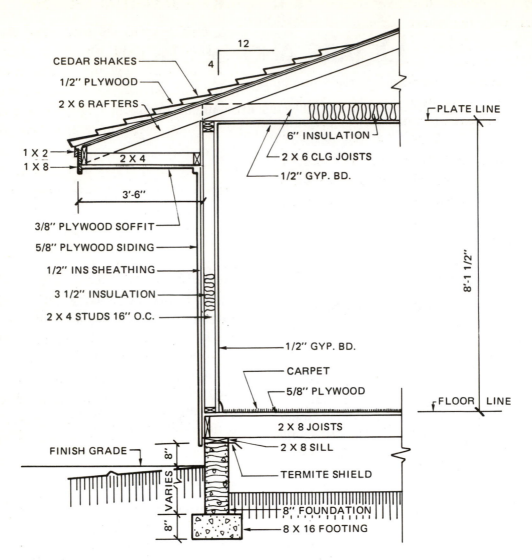

CEDAR SHAKES
1/2" PLYWOOD
2 X 6 RAFTERS
1 X 2
1 X 8
2 X 4
3'-6"
3/8" PLYWOOD SOFFIT
5/8" PLYWOOD SIDING
1/2" INS SHEATHING
3 1/2" INSULATION
2 X 4 STUDS 16" O.C.

12
4

6" INSULATION
2 X 6 CLG JOISTS
1/2" GYP. BD.

PLATE LINE

8'-1 1/2"

1/2" GYP. BD.
CARPET
5/8" PLYWOOD

FLOOR LINE

2 X 8 JOISTS
2 X 8 SILL
TERMITE SHIELD

FINISH GRADE
8"
VARIES
8"
8" FOUNDATION
8 X 16 FOOTING

Fig. 4-1 Typical wall section

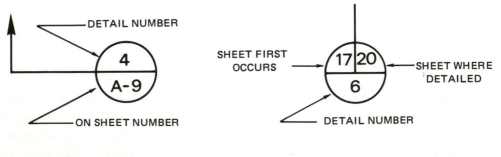

DETAIL NUMBER

4
A-9

ON SHEET NUMBER

(a)

SHEET FIRST
OCCURS

17 20
6

SHEET WHERE
DETAILED

DETAIL NUMBER

(b)

Fig. 4-2 Special section symbols

section, identifies the sheet where it first occurs and the sheet where the section drawing is located.

The location of some sections is so obvious that cutting plane lines are not required. The floor plan, a prime example, is really a horizontal section through the building, as shown in figures 3-8, 3-9 in Unit 3. The section labeled Typical Wall Section, figure 4-1, presents another example. This section could occur any number of places throughout the building, so its exact location is not important.

FULL SECTIONS

Full sections are sections through a whole building or object. Floor plans, as first discussed in Unit 3, are horizontal sections through the building. Two full sections cut vertically through the building are called *longitudinal sections* and *cross sections*. Longitudinal sections are taken lengthwise, and cross sections are taken across the width of the building. These sections are shown in figure 4-3, page 34.

Other types of the full section are the *design section* and the *structural section*. The purpose of the design section is to describe the inside of the building. The spaces occupied by the structural parts are left blank or else crosshatched. Footings and foundations are not shown in these design sections, figure 4-4, page 35. A structural section made at the same cutting plane is shown in figure 4-5, page 35. It shows the construction of foundation, floors, walls, and roof. Interior details usually are not shown, but it would be helpful if they were. In large buildings, when the size of the drawing prevents showing parts clearly in a full section, detail sections are required.

HALF AND PARTIAL SECTIONS

Half sections are used when the object is symmetrical and when it is desirable to show the inside and the outside in only one view. The half section is a clearer way to do this than the use of hidden lines even for a simple object, as in figure 4-6, page 36.

Partial sections show a special feature of an object without an extra view. Only a fraction of the object is shown in section thus retaining most of the character of the primary view, figure 4-7, page 36.

REVOLVED SECTIONS

A revolved section often explains the shape of parts by showing the shape in section directly on the object. Revolved sections are used to show cross-sectional shape of objects such as spokes, bars, and rods. The section is shown in place with the axes revolved 90°, figure 4-8, page 37.

A common case of the revolved section is the standard window and door frame detail. The jamb is revolved and appears between the head and sill. The revolved jamb is separated with break lines, in this case, because of the complexity of the detail. While the jamb is revolved, the head and sill are in their normal positions for a vertical section through the window or door, figures 4-9, 4-10, page 37.

SECTION CONVENTIONS

When a section detail is developed at an imaginary cutting plane, the outline of the cut material should be heavy. Other lines within the section should be thinner. All materials exposed in the cut section should be shaded or marked to identify the material. A few standard architectural material symbols

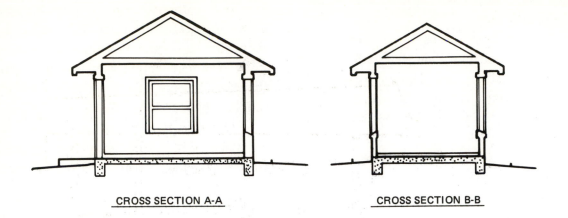

CROSS SECTION A-A   CROSS SECTION B-B

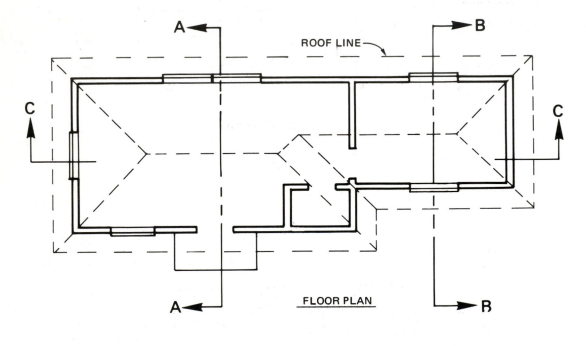

FLOOR PLAN

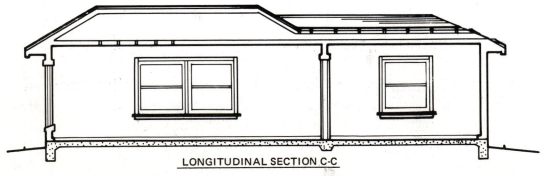

LONGITUDINAL SECTION C-C

Fig. 4-3 Longitudinal and cross sections

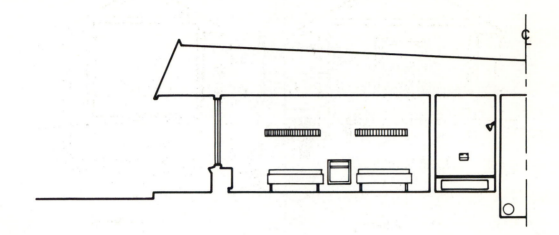

Fig. 4-4 Design section

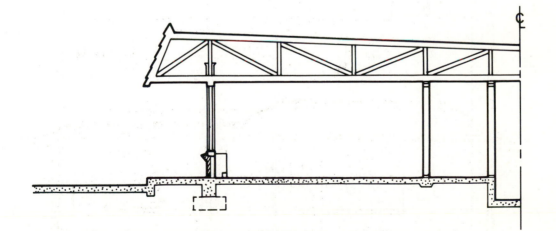

Fig. 4-5 Structural section

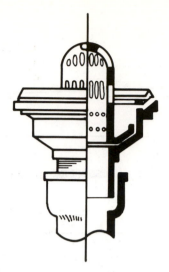

Fig. 4-6 Half section — roof drain

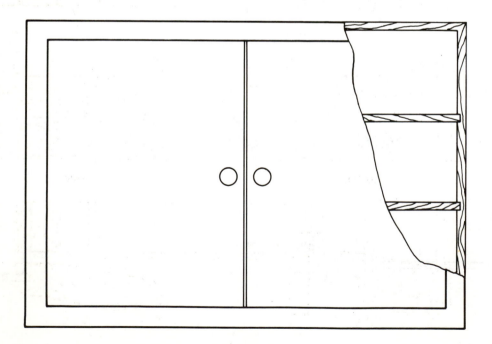

Fig. 4-7 Partial section

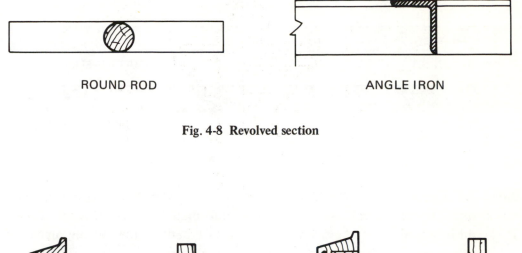

ROUND ROD   ANGLE IRON

Fig. 4-8 Revolved section

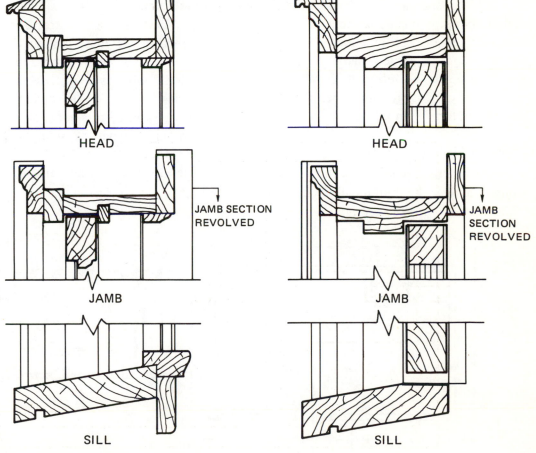

HEAD   HEAD

JAMB SECTION
REVOLVED

JAMB SECTION
REVOLVED

JAMB   JAMB

SILL   SILL

Fig. 4-9 Window sections   **Fig. 4-10 Door frame sections**

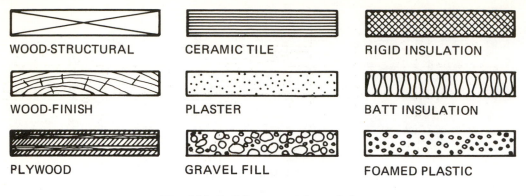

Fig. 4-11 Building materials symbols

are shown in figure 4-11. The appendix contains a complete list of symbols.

It should be noted that not every drafter uses all the standard material symbols presented here. When reading a particular blueprint, it is best to check the drawings for a legend or symbol list. Lacking this, it will be up to the reader to search the notes and specifications for the identification of unusual material symbols.

SPECIAL CASES

The cutting plane does not always run straight across the building. It may offset part way through to reveal more detailed information. The reader must determine the exact path of the cutting plane before trying to interpret the section, figure 4-12.

There are also exceptions to the rule to section all materials cut by the cutting plane. It is conventional practice not to section bolts, screws, rivets, shafts, rods, handles, etc. Even incidental parts like braces, spokes, ribs, etc. are not sectioned. They are drawn as side views within the parts that are sectioned. These parts are better explained when not sectioned, figure 4-13.

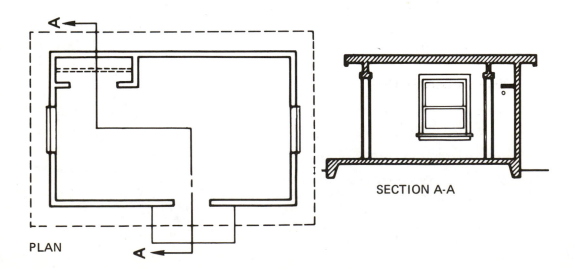

Fig. 4-12 Offset cutting plane

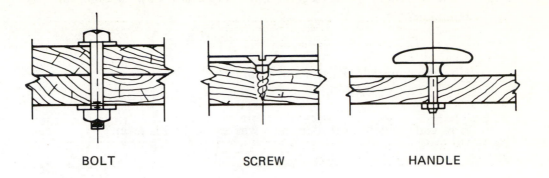

BOLT   SCREW   HANDLE

Fig. 4-13 Parts not sectioned

SUMMARY

- Details are enlarged views of portions of the building.
- Details may be sections, plans, or elevations.
- Sections are views of the interior construction of objects at imaginary cut planes.
- Sections are very important in blueprint reading.
- Cutting plane lines for section views usually are identified by special symbols.
- A floor plan is a horizontal section through the building.
- When cut vertically sections through buildings are either longitudinal sections or cross sections.
- Revolved sections are used to explain shapes directly on the objects.
- Not all drafters use standard material symbols uniformly.
- Section cutting planes sometimes offset part way through the object.
- Strict sectioning principles are not always followed when drawing minor parts.

REVIEW QUESTIONS

1. What is the term given to portions of drawings enlarged to give more information?

2. What do sections show better than other views?

3. What are the most important views in blueprints for buildings?

4. What is probably the second most important view in blueprints for buildings?

5. How are materials identified in sections?

6. How can a blueprint reader know where a section is taken?

7. Sketch a cutting plane symbol used when the section is drawn on the same sheet with the object.

8. Sketch a cutting plane symbol that is used when the section is drawn on another sheet.

9. How is a section labeled if no cutting plane symbol is used?

10. What are most full sections cut horizontally through the building known as?

11. Name a full section cut vertically through the building.

12. What is the purpose of the design section?

13. What is usually left out of design sections through a building?

14. What is the purpose of the structural section?

15. Name a section view that shows only part of the object cut.

16. Name an example of a common revolved section.

17. What drafting convention helps make sections easier to understand?

18. Which part of a section should be drawn darkest?

19. What is the purpose of an offset cutting plane line?

20. Name three items that are not sectioned when cut by the cutting plane.

21. Shown below are side views of five different shaped objects. Copy these views and sketch possible revolved sections directly on them.

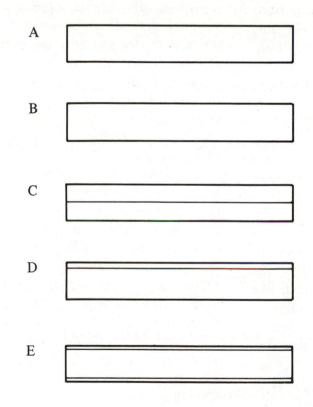

Unit 5
Pictorial Views

OBJECTIVES

After studying this unit, the student should be able to:

- explain the value of pictorial views compared to multi-view drawings.
- identify oblique projection drawings and give examples of their use.
- make oblique and isometric sketches of a simple object.
- name two types of perspective drawings.
- relate which pictorial drawings can be measured.

VALUE OF PICTORIAL VIEWS

The single plane views discussed in Units 3 and 4 often leave the viewer wondering about the true shape of the object. Even when the multi-view, single plane method is understood well, some shapes are difficult to visualize. Pictorial views are used to clarify these cases.

A good use for pictorial views is to show what the complete building looks like. Using suitable rendering techniques, the architect can judge the appearance of the design. The picture is used to show the owner and the public how the finished building will look. No special skill or training is required to understand this kind of drawing. It shows objects in three-dimensional form much like we actually see them. An understanding of the different types of pictorial drawings is of value because some of them can be measured even though others are useful only as pictures of the object. The three common types of pictorial drawings are the oblique, the iso-metric, and the perspective.

OBLIQUE VIEWS

Oblique-projection drawings are the simplest type of pictorial drawings. They have one face that is true size and shape, similar to one of the single plane, multi-view drawings discussed in Unit 3. This true face is usually the front side or the top side. An advantage of oblique drawing over other pictorial methods is that when the front view contains circles and arcs, the circles and arcs can be drawn with a compass. This type drawing is also suitable for objects whose intricate contours appear in only one face or direction.

Figure 5-1 illustrates three variations of oblique drawing. They differ only in the measurement of the third dimension or depth. Cavalier oblique drawing, although distorted in appearance, is the simplest to draw because all sides are drawn to full size, figure 5-1a. General oblique drawing, figure 5-1b, is the most realistic because it is most accurately proportioned. Cabinet oblique drawings are widely used but appear a little too shallow, figure 5-1c.

Figure 5-2 illustrates some applications of oblique drawing. Letters, exploded views, and objects with circles on or parallel to the front face are shown.

An oblique drawing with the top true shape and the height 3/4 size is another variation. The plan of a building can be traced and the vertical dimension added to make it pictorial. It is versatile in its application. Room partitions and furniture or equipment can also be shown. Examples of this type oblique drawing are shown in figure 5-3.

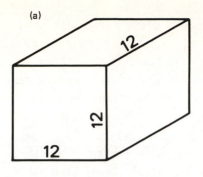
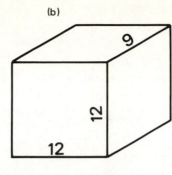
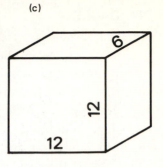

(a)

12

12

12

(b)

9

12

12

(c)

6

12

12

CAVALIER OBLIQUE
QUICKEST TO DRAW

GENERAL OBLIQUE
BEST PROPORTIONS

CABINET OBLIQUE
MOST USED

Fig. 5-1 Types of oblique drawings

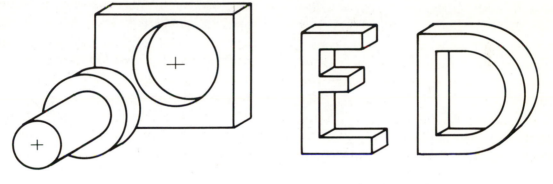

EXPLODED DRAWING

SHALLOW LETTERS

Fig. 5-2 Oblique drawing examples

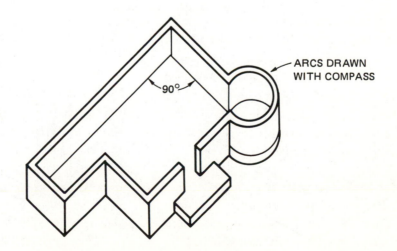

ARCS DRAWN
WITH COMPASS

90°

Fig. 5-3 True shape oblique plan method

ISOMETRIC VIEWS

Isometric drawings often are used for special details or assemblies in building plans. They show a three dimension picture in one drawing. Most drafters seem to prefer isometric drawings. Although somewhat distorted when compared to a perspective, they are quite simple to draw and to understand. The object is viewed from one corner and from above. None of the surfaces is shown in true shape, but measurements in all three planes are full sized. Figure 5-4 shows the basic construction of the isometric drawing. Other examples of simple objects are shown in figure 5-5.

Builders often have difficulty visualizing details when working from floor plans to elevations and back. Sometimes it is helpful for the builder to make isometric drawings to coordinate information from several sheets of plans.

There are several common uses for the isometric drawing method. Heating and air-conditioning duct work can be shown to advantage. Lengths of piping or ducts can be measured directly from these drawings. They also help the builder visualize how the pipes and ducts fit the building, figure 5-6. Another common use of the isometric drawing is the plumbing riser detail, figure 5-7.

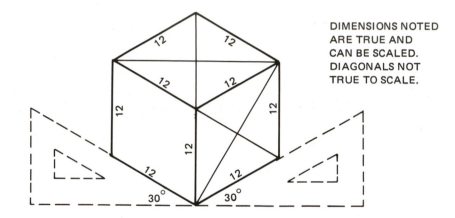

DIMENSIONS NOTED
ARE TRUE AND
CAN BE SCALED.
DIAGONALS NOT
TRUE TO SCALE.

Fig. 5-4 Isometric construction

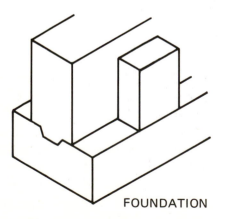

FOUNDATION

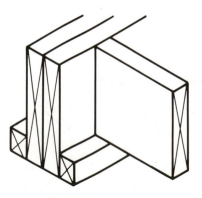

GIRDER DETAIL

Fig. 5-5 Isometric examples

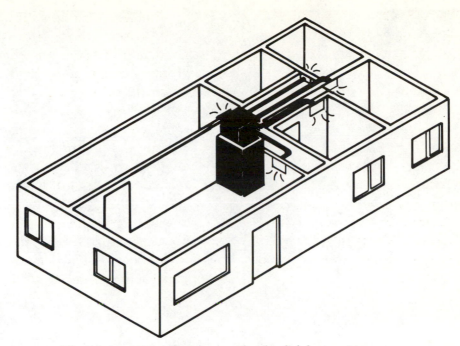

Fig. 5-6 Isometric view of an overhead radial duct system

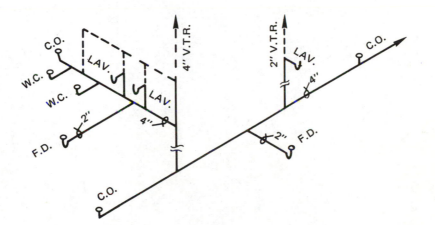

Fig. 5-7 Isometric plumbing riser

PERSPECTIVE VIEWS

Perspective drawing is the most realistic pictorial method of representing a building or object. It presents the building very much as would a photograph. A perspective view of the proposed building sometimes is printed on the cover of the blueprints, figure 5-8, page 46. This is very helpful to the builder in getting acquainted with the basic design.

Perspectives printed in blueprints sometimes are simplified or stylized, but still quite easy to understand.

More precise and natural looking perspectives are rendered in color. They are framed for display and used by the architect and the owner to promote the new building. It is interesting to compare a photograph of the finished building with the architect's perspective, figures 5-9, 5-10, page 47.

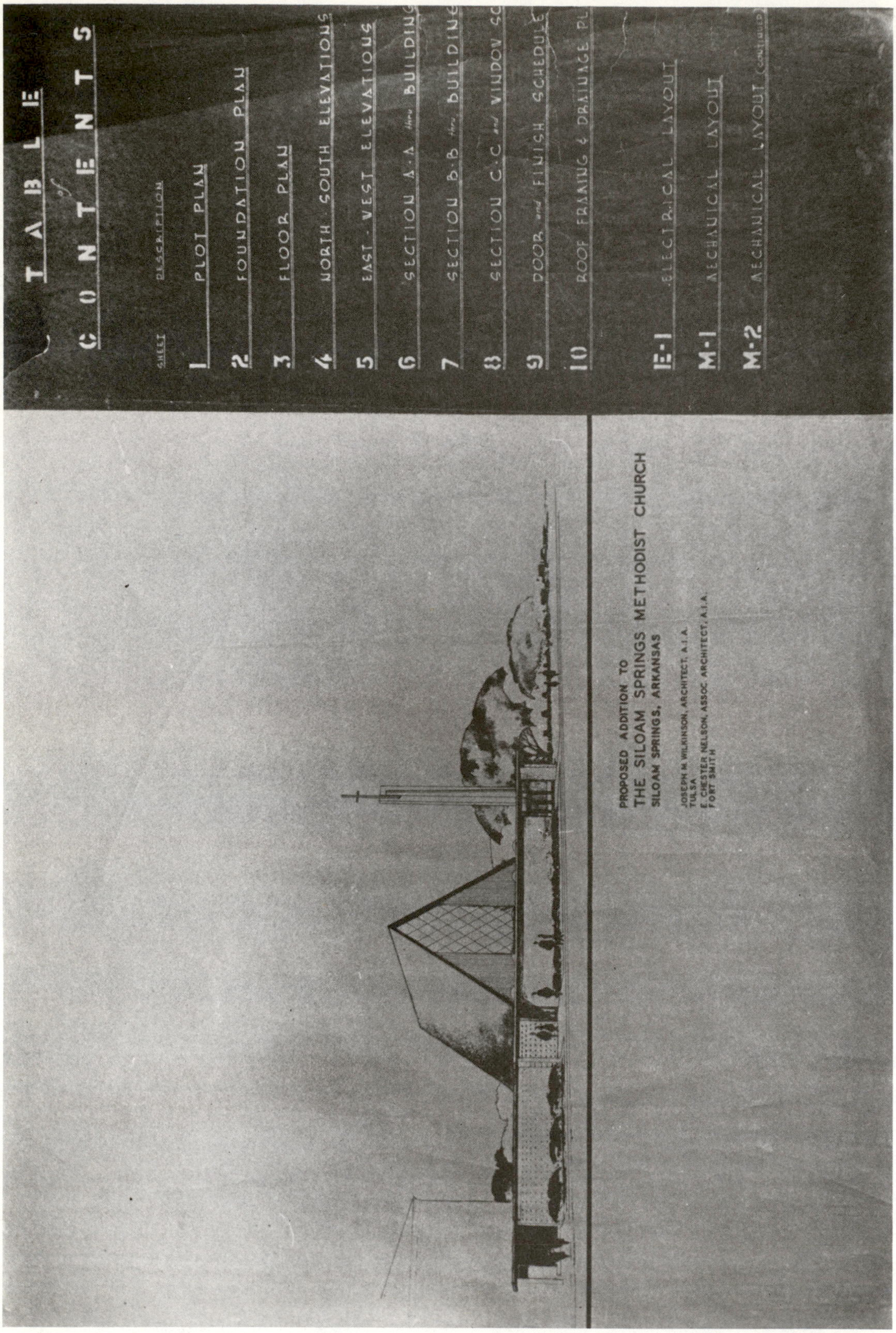

Fig. 5-8 Perspective of building on blueprints

Fig. 5-9 Perspective of proposed building

Fig. 5-10 Photograph of completed building

TWO-POINT PERSPECTIVE

Two-point perspective is the type most often used. The object is turned slightly so that two sides are in view. If the horizontal lines of a side are extended, they converge (meet at one point). This point is called the *vanishing point.* With two sides showing, there are two vanishing points, hence the name two-point perspective. The vanishing points may be some distance to the right and left of the drawing, figure 5-11, page 48. The vertical lines are drawn parallel to each other and do not converge.

ONE-POINT PERSPECTIVE

Another common type is the one-point perspective. Here the object is viewed straight on. The converging lines of receding planes all meet in one point somewhere on the perspective drawing. In certain cases it seems more effective and more convenient to use than the two-point method. One-point perspectives are the best choice for interior views, figure 5-12, page 48. They also can be used for some exterior views such as entrance details and symmetrical groupings of buildings, figure 5-13, page 48, and figure 5-14, page 49.

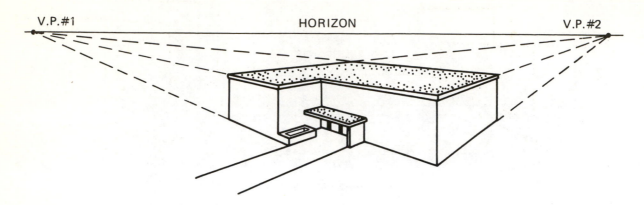

Fig. 5-11 Two-point perspective

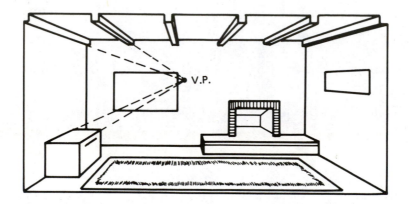

Fig. 5-12 One-point interior perspective

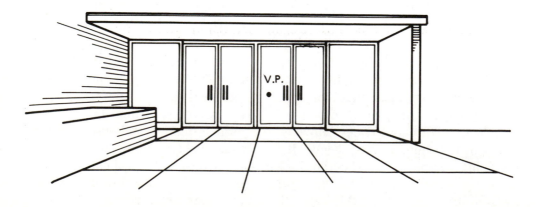

Fig. 5-13 One-point entrance perspective

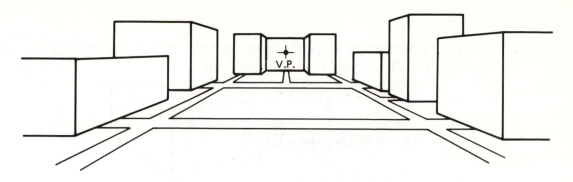

Fig. 5-14 One-point exterior perspective

IDENTIFICATION CUSTOM

Most drafters make a notification on the plans of the type of view used. This is true especially of isometric details. The builder, knowing that a view is an isometric, can measure the parts with confidence. Likewise, recognizing a perspective, the builder does not try to get any useful measurements from the drawing but simply relates the plans to the finished building form.

SUMMARY

- Pictorial drawings help the viewer visualize the forms drawn in multi-view drawings.

- Pictorial views help the architect judge the design.

- Pictorial views are easy to understand because they look much like a photograph of the object.

- Oblique drawings have one face that is true size and shape.

- Intricate shapes and circles are easier to draw by using oblique views than by using other pictorial methods.

- Isometric drawings are preferred by most drafters when making pictorial details.

- Measurements are accurate in all three planes of an isometric drawing.

- Perspectives are the most realistic pictorial drawings.

- Perspectives rendered in color are used to promote new buildings.

- Two-point perspectives are usually used for exterior views of a building.

- One-point perspectives are best used for interior views.

- Drafters usually identify isometric and perspective drawings on the plans.

REVIEW QUESTIONS

1. What kind of drawings are easier to understand than multi-view drawings?

2. Why is no special skill or training needed to understand pictorial views?

3. Why is it important to know the different types of pictorial drawings if no special skill is needed to understand them?

4. The simplest type of pictorial drawing has one face drawn true size and shape. What is this type called?

5. How can floor plans be made three-dimensional using the oblique method?

6. Isometric drawings show three sides of an object. How many sides are true shape?

7. Which dimensions in an isometric drawing are true length and can be measured?

8. Name a common use of isometric drawings in building plans.

9. Which type pictorial drawing is most realistic?

10. Why do architects make color renderings in perspective of proposed buildings?

11. When the object is turned so that two sides are seen, what is the perspective view called?

12. When the object is viewed straight on, how many vanishing points are used to draw the perspective?

13. What type perspective is usually used for exterior views of a building?

14. Which lines or dimensions on perspective views can be measured for useful information?

15. Sketch the object shown below:

    a. in oblique projection.

    b. in isometric projection.

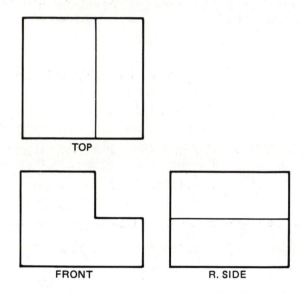

TOP

FRONT   R. SIDE

Unit 6
Dimensioning

OBJECTIVES

After studying this unit, the student should be able to:

- accurately measure objects drawn to scale.
- explain the difference between the architect's and the engineer's scales.
- give wood frame dimensioning conventions.
- give masonry building dimensioning conventions.
- explain the use of modular coordination in building plans.

SCALE

After the graphic principles of Units 2 through 5 have been applied, the next step is to determine the size of the object. Drawings in blueprints for buildings rarely are drawn full-size. Building parts are generally too large to show full-size on paper. Instead, the drawings are made smaller but still proportional to the real size. They are drawn to *scale*. A convenient scale is selected and used to proportion all distances shown on the drawing. The scale used by the drafter is noted on the drawing. It may be part of the view title as in figure 6-1, or it may be given in the title

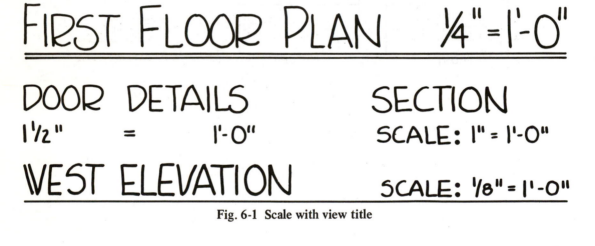

FIRST FLOOR PLAN   ¼"=1'-0"

DOOR DETAILS
1½"   =   1'-0"

SECTION
SCALE: 1"=1'-0"

WEST ELEVATION

SCALE: ⅛"=1'-0"

Fig. 6-1 Scale with view title

JOB NUMBER:	
DATE:	TOWER A
DRAWN BY:	3250 NORTH
CHECKED BY:	ALLENDALE
SCALE: ¼"=1'-0"	
REVISED:	FIRST FLOOR

Fig. 6-2 Scale in title box

box for the drawing, figure 6-2. Where several different scales are used on the same sheet, the title box may say "SCALE AS NOTED." In any case, the builder must determine the scale of the drawing before attempting to measure anything on it.

At this point, a distinction of the term "scale" must be emphasized. In the previous remarks, scale referred to the actual measurements that made an object smaller but still proportional to real size. *Scale* also refers to certain "drawing instruments" used by architects and engineers for the purpose of measuring "to scale."

ARCHITECT'S SCALE

Most of the measurements and scales on the drawings are done using the architect's

scale. Sections of different shaped drafting scales are shown in figure 6-3. While experienced drafters often prefer the flat scale, the triangular scale has eleven different scales and is the best all around choice for blueprint reading. A list of the scale measurements is given in figure 6-4.

The architect's scale is open divided. Open-divided scales have the main units undivided, with a fully subdivided extra unit placed at the zero end of the scale. Except for one full-size scale, two scales are combined on each edge. The combined scales are compatible; one is twice as large as the other. Their zero points are on opposite ends of the scale so one scale is read left to right and the other one right to left, as shown in figure 6-5, page 54.

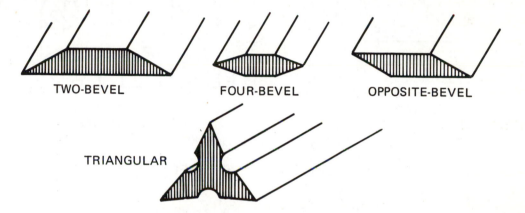

TWO-BEVEL   FOUR-BEVEL   OPPOSITE-BEVEL

TRIANGULAR

Fig. 6-3 Typical drafting scale shapes

SIDE 1	FULL SIZE	. . . .	1/16" GRADUATIONS	
	3/32" = 1' - 0"	. . . .	1" = 1' - 0"	
SIDE 2	1/2" = 1' - 0"	. . . .	1" = 1' - 0"	
	1/8" = 1' - 0"	. . . .	1/4" = 1' - 0"	
SIDE 3	3/8" = 1' - 0"	. . . .	3/4" = 1' - 0"	
	1 1/2" = 1' - 0"	. . . .	3" = 1' - 0"	

Fig. 6-4 Architectural scales

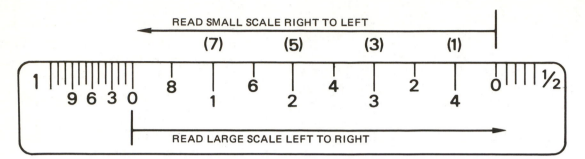

6" ARCHITECT'S SCALE

Fig. 6-5 Open-divided combined scales

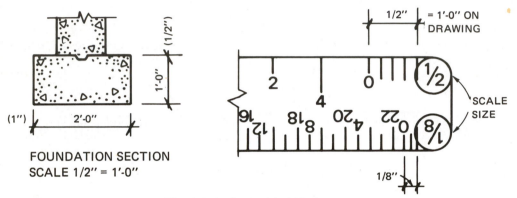

FOUNDATION SECTION
SCALE 1/2" = 1'-0"

Fig. 6-6 Scale size identified

In architectural drawing, the major unit of measurement is the foot. The subdivided extra unit at the zero end divides the foot into inches as well as fractions of inches on the larger scales. The fraction or number near the zero end of each scale identifies the scale. This number indicates the unit length of the scale in inches and represents one foot on the actual building, figure 6-6.

To measure correctly with a scale, place the proper edge parallel to the line to be measured. Determine the number of whole feet in the line's scale distance. Index the scale with this number placed on one end of the line. The other end will fall on zero or slightly beyond in the subdivided extra unit space. The length of the line is now read as so many feet and inches, figure 6-7.

ENGINEER'S SCALE

Occasionally an engineer's scale is used by architectural drafters for drawing land measurements, plot plans, and stress diagrams. Engineer's scales are always fully divided and can be obtained in any of the flat or triangular types, figure 6-3. The six scales found on the triangular type are: 1" = 10', 1" = 20', 1" = 30', 1" = 40', 1" = 50', 1" = 60'.

Land measurement is always given in feet and decimal parts of a foot rather than in inches. One inch on the 1" = 10' scale would have 10 divisions. One inch on the 1" = 20' scale would have 20 divisions. Decimal parts of a foot are estimated. The scale is identified by a number near its zero end. This number is the number of parts per inch into which the scale is divided. Except for the 1" = 10'

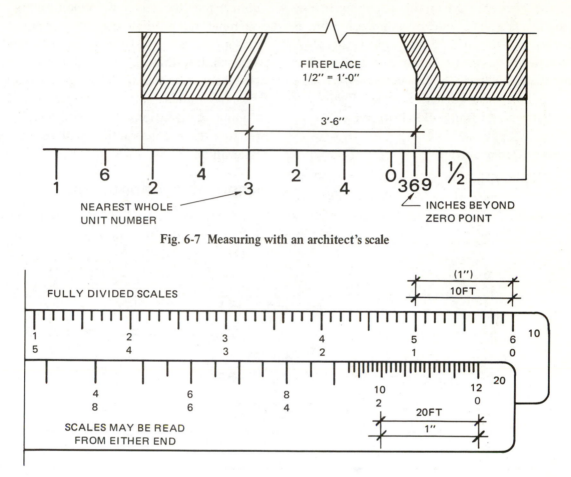

Fig. 6-7 Measuring with an architect's scale

Fig. 6-8 Engineer's scales

scale, divisions on the scales are labeled 0, 2, 4, 6, 8, etc., the zero being dropped for simplicity. The 1″ = 10′ scale has every ten feet labeled as 1, 2, 3, 4, etc., figure 6-8.

WRITTEN DIMENSIONS

As important as the scale of a drawing is, the written dimensions are even more important. Scaled dimensions should never be used if written dimensions are given. Written dimensions are more accurate. Blueprint paper usually shrinks slightly making precise scaling impossible. Also, the drafter may on occasion change the dimension of a part without redrawing it. Without accurate dimensioning, many expensive mistakes can occur.

DIMENSIONING CONVENTIONS
Frame Construction

Wood frame buildings are dimensioned from the exterior face of the studs in the outside walls. This rule applies regardless of the kind of finish used, such as brick or stone veneer. Window and door openings are located by their center lines. Interior walls are located also by dimensioning their center lines. Sometimes partition dimensions are placed outside the plan, but many times they run across the plan. Overall dimensions are the farthest out from the object. Building offsets and major structural features are next. Window and door dimension lines are the nearest to the object. The purpose is to avoid having dimensioned spaces cross extension lines of shorter elements. Careful attention to the arrowheads at each end of a dimension line is necessary to be sure of just what is included. When there is doubt because of a missing or obscure arrowhead, the labeled dimension line should be measured with a scale. Figure 6-9 illustrates wood frame dimensioning.

Masonry Construction

Masonry buildings are dimensioned to the extreme face of their walls, both exterior and interior. These are more precise locations than in frame construction and usually represent finished surfaces. Windows and doors are located by dimensioning their openings rather than their center lines. The masonry opening usually is given by the door and window manufacturer, and the architect fits them into the wall. (Figure 6-10 shows masonry dimensioning.) Other special dimensioning conventions for commercial fire-resistive type construction will be treated in later units.

MODULAR COORDINATION

Modular Coordination is the name given to the effort to standardize the size and

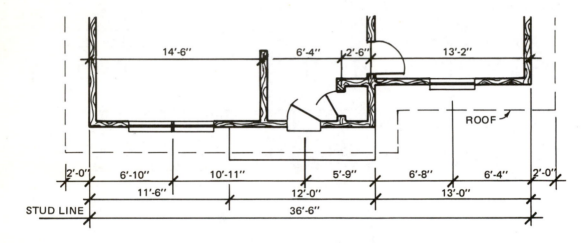

Fig. 6-9 Wood frame dimensioning

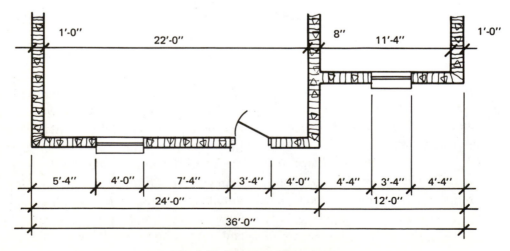

Fig. 6-10 Masonry dimensioning

assembly of building materials. Modular construction methods result in simplification of details which in turn result in economy of labor and materials throughout the building industry.

International standards activities have established the 4″ basic module for countries using the English system and 10 cm for countries using the metric system. A *module* is a unit of measurement used to describe or control building elements. The 4″ module is small enough for unit masonry materials and is readily adapted into a number of larger planning modules. Planning modules of 2′-8″, 3′-0″, 3′-4″, 4′-0″, 5′-4″, etc. are all multiples of the universal 4″ module.

The architect selects a suitable planning module and locates major structural elements on a grid based on the module selected. It is understood that a 4″ subgrid is included though not shown except in large scale details. The 4″ subgrid is three-dimensional, controlling both horizontal and vertical construction details, figure 6-11.

Manufacturers develop many modular products for the building industry. Masonry products are sized so that the distance from center line to center line of joints is some multiple of 4″. Grid lines are located at these center-of-joint points, figure 6-12. Larger

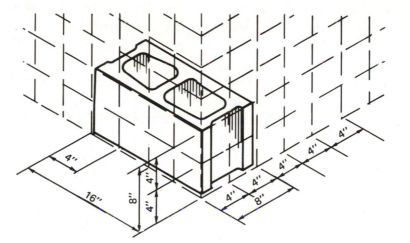

Fig. 6-11 4″ Three-dimensional grid

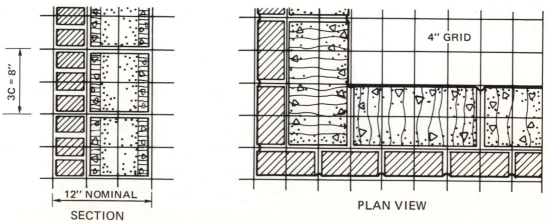

Fig. 6-12 Modular masonry on grid

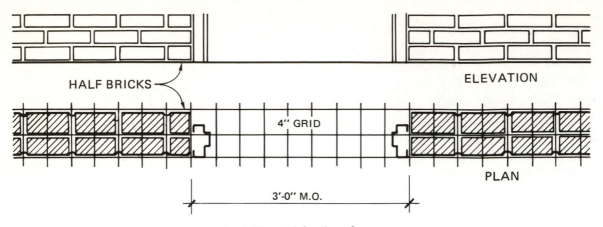

Fig. 6-13 Modular door frame

components, such as door frames, are sized to fit between modular masonry units without extra cutting, figure 6-13. In any case, significant simplification is effected in the dimensioning of modular blueprints.

The layout of a floor plan must be done with care to keep the dimensions in proper form. Modular blueprints are usually identified with a note stating that the drawings are modular. In order to standardize the modular note, the American Institute of Architects has devised a translucent appliqué which is affixed to the original drawing. It is recommended by the Modular Measure Committee of the American Standards Association and is usually placed on the first drawing, figure 6-14.

Special instructions are given in this modular note to assist the builders in interpreting the plans. Of special significance is the use of arrowheads for dimensions that end on a 4″ grid line. Round dots are used for dimensions off the grid lines. Note that modular masonry is dimensioned to the center line of mortar joints. A 12″ brick wall is actually 11 5/8″ thick. A 2 x 4 stud wall is also dimensioned centered in a 4″ module, figure 6-15, page 60.

Not all blueprints with structural grids follow these modular drafting standards. The architect may begin a design by using a grid but then choose to use more common dimensioning methods. This is not a big problem to the blueprint reader if the concept of modular coordination is understood. Any effort on the part of the designer to better organize the design and the dimensions that describe it are welcomed.

SUMMARY

- Blueprints of buildings are made at reduced size, scaled to fit on the sheets.
- The scale used to make the drawing is always noted somewhere on the sheet.
- Most building plans are drawn using the architect's scales.
- Architect's scales are open divided with each unit representing one foot.
- Engineer's scales are used for land measurements, plot plans, and stress diagrams.

NOTE—All drawings are dimensioned by The Modular Method

in conformance with the American Standard Basis for Coordination of Dimensions of Building Materials and Equipment, A62.1

This system of dimensioning is used for greater efficiency in construction: less cutting, fitting and waste of material, less chance for dimensional errors. The Modular Method uses a horizontal and vertical grid of reference lines. The gridlines are spaced 4 inches apart in length, width and height.

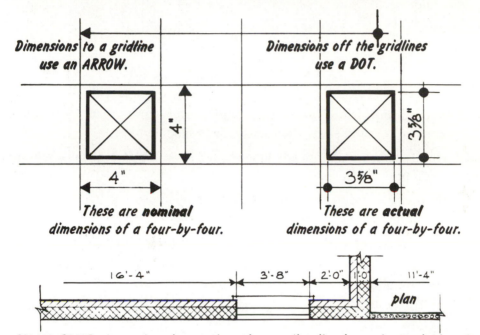

Dimensions to a gridline use an ARROW.

4"

4"

These are nominal dimensions of a four-by-four.

Dimensions off the gridlines use a DOT.

3⅝"

3⅝"

These are actual dimensions of a four-by-four.

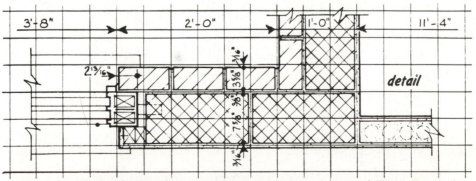

16'-4"   3'-8"   2'-0"   1'-0"   11'-4"

plan

*SMALL-SCALE plans, elevations and sections ordinarily give only nominal and grid dimensions (from gridline to gridline in multiples of four inches, using arrows at both ends). Dimension-arrows thus indicate **nominal** faces of walls, jambs, etc., finish floor, etc., coinciding with invisible gridlines, which are not drawn in at such small scales.*

3'-8"   2'-0"   1'-0"   11'-4"

2'9/16"

detail

*LARGE-SCALE detail drawings actually show these same gridlines drawn in, every 4 inches. On these details, reference dimensions give the locations of **actual** faces of materials in relation to the grid.*

Fig. 6-14 Appliqué that may be attached to working drawings to indicate they are modular

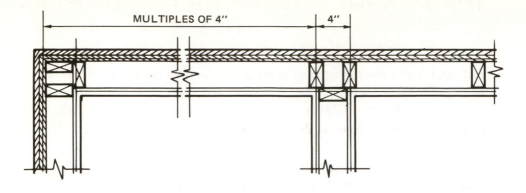

Fig. 6-15 Modular layout of 2 x 4 walls

- Written dimensions on a drawing are more accurate than scaled dimensions and should always be used when available.

- Wood framed buildings are dimensioned from the face of the studs even when brick veneered.

- In frame buildings, window and door openings are located by their center lines.

- Wood framed partitions also are located by their center lines.

- Masonry buildings are dimensioned to the face of their walls, both exterior and interior.

- In masonry construction, windows and doors are located by dimensioning their openings.

- Modular coordination is the name given to the effort to standardize the size and assembly of building materials.

- The basic module for the United States is 4″.

- Planning modules are multiples of 4″ and are used by the architect to locate main parts of the design.

- The 4″ module is three-dimensional.

- Dimensions ending on 4″ module grid lines are marked with arrowheads while those ending off the grid are marked with dots.

- Many buildings are drawn with planning grids without using modular dimensioning standards.

REVIEW QUESTIONS

1. Why must building parts be drawn to scale rather than full size on blueprints?

2. When several different scales are used on the same sheet, what note might be found in the title box?

3. Name the two types of scales used to draw building plans.

4. What is meant when a scale is open divided?

5. What is the major unit of measurement on architectural scales?

6. How are dimensions less than a foot scaled on open divided scales?

7. Give a typical scale used for building plans and tell how it is found on the architect's scale.

8. How are the engineer's scale units laid out, open or fully divided?

9. Name a use for engineer's scales in building plans.

10. Identify two engineer's scales.

11. When are scaled dimensions taken off the plans?

12. Give one reason why written dimensions are better than scaled dimensions.

13. In wood frame construction exterior walls are dimensioned to the exterior face of the studs. How are partitions dimensioned?

14. How are windows and doors dimensioned in wood frame construction?

15. How are masonry walls dimensioned on floor plans?

16. How are windows and doors in masonry walls dimensioned?

17. What is the basic module adopted in this country for building materials?

18. What special arrowhead standards are used in modular dimensioning?

19. Where is modular masonry dimensioned?

20. Using an architect's scale, measure the following lines.

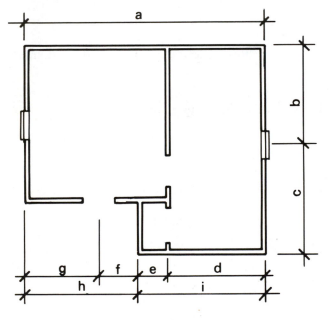

FLOOR PLAN   1/8″ = 1′-0″

SECTION II
CONSTRUCTION CONTRACT DOCUMENTS

UNIT 7
SPECIFICATIONS

OBJECTIVES

After studying this unit, the student should be able to:

- list the items included in a construction contract.
- explain the purpose of the general conditions section of the specifications.
- discuss the scope of the Construction Specification Institute's specification format.

CONSTRUCTION CONTRACTING

Most workers involved in commercial construction in this country work for general contractors. *General contractors* are independent contractors who build under a contract and do 15% or more of the work with their own work force. Building under a contract means that an agreement has been made to do a job for someone for a fixed amount of money. When signed by both parties, this contract is a legal matter and is enforceable by law.

A *construction contract* is more than a simple piece of paper signed by the contractor and the owner. It includes the agreement (which is signed), the specifications (general and technical), and the drawings (working drawings and shop drawings).

SPECIFICATIONS — GENERAL

The specifications are the written instructions prepared by the architect for constructing a building. In most cases they are bound in book form. As noted above, one section of the specifications is general in nature. This section is called the *general conditions*. It is usually the first section of the contract and

defines the relationship among the agreeing parties as well as their responsibilities. It tells how the contractor is paid and when. Insurance, changes in the work, and correction of faulty work are other topics included.

Of special interest to the blueprint reader is the relationship between the specifications and the drawings. The information given in the specifications should reflect information given in the drawings. If a difference occurs, it is better to follow the directions given in the specifications. The drawings take much more time to produce, and every item is not always checked with the final specifications. Usually the specifications are prepared last and reflect final decisions. To avoid the problem, architects usually include information about specific kinds of material in the specifications only, while the drawings show type, size, and location of material.

CHANGE ORDERS

It is provided for in the general directions that the owner may order changes in the work such as additions, deletions, or other revisions. A *change order* is a written order to the contractor authorizing a change in the work. When properly done; it becomes part of the contract and must be followed in the work. Many change orders originate with the architect to take care of unforseen conditions. Some change orders involve revising drawings as well as written specifications. A sample change order is shown in figure 7-1.

SPECIFICATIONS — TECHNICAL

Strictly speaking, the general section of the specifications deals with contractual legal matters and is not really specifications. Historically it has been bound with the technical specifications, and this entire collection of information was called the "specifications." A better name for this printed material is *Project Manual.* When this term is used, specifications perform their real function of describing the technical requirements of the construction.

A set of specifications is divided into sections so that each trade or type of work

---

CHANGE ORDER

No. 1

Date 1-6-77

Project: TOWERS APARTMENTS
3250 North Oak
Allendale, Nebraska

To: Eagle Construction Co.
2900 Bluff Road
Canyon City, Oklahoma

Revised Contract Amount

Previous contract	$1,525,050.00
Amount of this order (~~decrease~~) (increase)	15,100.00
Revised contract	$1,540,150.00

The contract time is hereby (increased) (~~decreased~~) (~~unchanged~~) by ___15___ days.

This order covers the contract revision described below:

Install ceramic mosaic floor tile in all bathrooms in place of vinyl floor covering specified. Tile shall be standard grade TCA 137.1 set with Dry-Set mortar ANSI A118.1. Installation shall conform to ANSI A108.5 for this method.

The work covered by this order shall be performed under and be part of the original construction contract.

Changes Approved

_____
Owner

by _____
Contractor

Brown and Smith, Architects

by _____

Fig. 7-1 A change order

```
    1.  General Conditions
    2.  Excavation
    3.  Concrete
    4.  Unit Masonry
    5.  Carpentry
    6.  Roofing and Sheet Metal
    7.  Dry-wall Construction
    8.  Ceramic Tile
    9.  Millwork
   10.  Painting and Decorating
   11.  Plumbing and Heating
   12.  Electrical
```

Fig. 7-2 Specification sections — small project

THE C. S. I. FORMAT FOR SPECIFICATIONS

DIVISION 1: GENERAL REQUIREMENTS Administrative and technical provisions which may apply to more than one Division.

DIVISION 2: SITE WORK Includes most subjects dealing with site work from subsurface exploration through landscaping.

DIVISION 3: CONCRETE Includes formwork, reinforcement, finishes, precast concrete, poured concrete, poured gypsum decks.

DIVISION 4: MASONRY Includes most materials traditionally installed by masons.

DIVISION 5: METALS Includes structural steel, metal decking and ornamental metal. Reinforcing steel for concrete is not included.

DIVISION 6: WOOD AND PLASTICS Includes rough carpentry, heavy timber construction, finish carpentry, architectural woodwork, and structural plastic work. No concrete formwork.

DIVISION 7: THERMAL AND MOISTURE PROTECTION Includes waterproofing, insulation, roofing of all types, sheet metal, and flashing.

DIVISION 8: DOORS AND WINDOWS Includes metal doors, frames, wood and plastic doors, store-front work, windows, hardware, and glazing.

DIVISION 9: FINISHES Includes interior finish materials, painting, and wall covering.

DIVISION 10: SPECIALTIES Includes special miscellaneous items such as chalkboards, fireplaces, flagpoles, lockers, and movable partitions.

DIVISION 11: EQUIPMENT Includes special equipment installed in the building.

DIVISION 12: FURNISHINGS Includes artwork, drapery, furniture, and seating.

DIVISION 13: SPECIAL CONSTRUCTION Includes air supported structures, clean rooms, incinerators, special purpose rooms, vaults, and swimming pools.

DIVISION 14: CONVEYING SYSTEMS Includes dumbwaiters, elevators, moving stairs, and pneumatic tube systems.

DIVISION 15: MECHANICAL Includes most items generally associated with plumbing, heating, and air conditioning.

DIVISION 16: ELECTRICAL Includes most items generally associated with the electrical trades.

Fig. 7-3 The C. S. I. format for specifications

is conveniently located. It is also good to list the sections in the same sequence as they occur on the job. For example, the carpentry section follows the concrete and comes before the roofing section. This simplifies the task of finding the information needed by each trade. Not all jobs have the same requirements; therefore, the list of sections varies. Nor does their order always fit the job sequence used by the contractor. Figure 7-2, page 65, shows the arrangement of typical sections for small jobs.

C. S. I. FORMAT

For a number of years, the Construction Specifications Institute (C. S. I.) has been developing a standardized specification format. A *format* is a special way of doing something. The C. S. I. format for specifications is based upon the "Division-Section" concept. The sections denote trade or basic unit of work. The divisions arrange the similar sections into logical trade work groups. The system provides for basic major divisions, forming the permanent backbone of the format. The sections, forming the flexible bits of information, are the main substance of the specifications.

The C. S. I. format is adaptable to small jobs but finds its greatest use in large commercial projects. There are 16 divisions in the C. S. I. format. They are given in figure 7-3, page 65, with a brief description of the content of each.

There is also a C. S. I. 3-Part-Section Format now adopted by most specification writers. The purpose of this format is to group the instructions in a section in the same order each time. If this is done, the reader knows just where to look for the information needed, thus making technical specifications much easier to use. The 3-Part-Section Format is described in figure 7-4. An example of its use is shown in figure 7-5.

Part 1 — General	Part 2 — Products	Part 3 — Execution
Description	Materials	Inspection
Quality	Mixes	Preparation
Assurance	Fabrication and	Installation
Submittals	Manufacture	Application
Product Delivery		Performance
Storage and		Erection
Handling		Field Quality
Job Conditions		Control
Alternatives and		Adjust and Clean
Allowances		Schedules
Guarantee		

Fig. 7-4 C. S. I. 3-part-section format

SECTION 06100 ROUGH CARPENTRY

1. GENERAL

    1.01 DESCRIPTION

    A. Work included: all wood framing indicated on the drawings.

    B. Related work not included: concrete formwork, wood doors and frames.

    1.02 QUALITY ASSURANCE

    A. Provide skilled workmen thoroughly familiar with the type framing included.

    B. Grading by rules of Southern Forest Products Association and American Softwood Lumber Standard (PS 20-70)

    1.03 PRODUCT HANDLING

    A. Store all materials safe from damage due to weather or dampness.

    B. Keep all grade marks legible.

2. PRODUCTS

    2.01 MATERIALS

    A. Plates and blocking — Utility Light Framing

    B. Studs   — Stud grade in 2 x 4 sizes. No. 2 in 2 x 6

    C. Joists   — No. 2 MG

    D. Beams & Girders   — No. 1

    E. Sheathing   — AD Fir Plywood

3. EXECUTION

    3.01 INSPECTION

    A. Verify all work complete and ready for framing before beginning this section.

    B. In case of discrepancy, notify the architect and postpone start of work.

    3.02 WORKMANSHIP

    A. All joists shall be made true, tight, and well nailed.

    B. Select members so that knots and obvious defects will not occur at joists.

    C. Discard lumber rejected by the architect because of excessive warp, twist, bow, crook, or mold, as well as improper cutting.

    3.03 QUALITY CONTROL

    A. Align framing for finish wall surfaces to vary not more than 1/8 inch from the plane of adjacent members.

    3.04 CLEANING UP

    A. Keep work area in a neat, safe, and orderly condition, free from accumulation of sawdust, cut-ends, and debris.

    B. At the completion of this portion of the work, thoroughly broom clean all surfaces.

Fig. 7-5 Example of the use of the C. S. I. 3-part-section format

SUMMARY

- Most commercial trades people work for general contractors.

- General contractors build by contract using their own forces for at least 15% of the work.

- A construction contract includes the agreement, the specifications, and the drawings.

- Specifications are the written instructions for construction work.

- The General Conditions contain rules to guide the parties during the contract.

- The specifications take precedence over the drawings.

- Change orders outline additions or deletions in the work and become part of the contract.

- Technical specifications are divided into sections, each describing some phase of the work.

- The Construction Specifications Institute has developed a 16-division specification format.

- The C. S. I. specification format is used in most commercial construction contracts.

- Most specification writers have adopted the C. S. I. 3-Part-Section Format for writing technical specifications.

REVIEW QUESTIONS

1. For whom do most construction workers work?

2. What are the three main parts of a construction contract?

3. Which parts of a construction contract are most important to the working trades?

4. What are specifications?

5. What are the two types of specifications?

6. If there is a conflict between what the drawings say and what the specifications call for, which must be followed?

7. What is the name of the document used to alter the work after the contract is in force?

8. With what do general specifications mainly deal?

9. What is a better name for the written materials usually called specifications?

10. In what order are the sections of technical specifications arranged?

11. What are the major parts of the C. S. I. format for specifications called?

12. How many divisions are found in the C. S. I. format?

13. How are the sections organized in the C. S. I. system?

14. In what division is information about concrete formwork found?

15. Where is rough and finish carpentry work specified in the C. S. I. format?

Unit 8
Working Drawings

OBJECTIVES

After studying this unit, the student should be able to:

- explain two important purposes of working drawings.
- give the usual arrangement of drawings in a set for small projects.
- tell how plans for large commercial structures are organized.

IMPORTANCE OF WORKING DRAWINGS

Working drawings (blueprints) make up a major part of the drawings prepared for the construction of a building. They are drawn by the architect during the planning stage. They are used first by the builder in preparing an estimate of the cost of the work when bids are placed for the job. Being on a par with the specifications, they are a major part of the contract. After the contract is awarded to the low bidder, they are used to build the building.

ARRANGEMENT OF DRAWINGS ON SHEETS

The preparation of the working drawings is the most time consuming part of the architect's work. Efforts to simplify and standardize this process result in certain drafting conventions listed in earlier units. Another standardization is the arrangement of the drawings on the sheets. The sheet size is based on the building size and scale of the drawing. Plans and elevations of commercial buildings usually are drawn at 1/8" = 1'-0". It is typical to draw only one plan view per sheet. If there is more than one floor to the building, each floor plan is drawn on a separate sheet. This simplifies the drafting process as one floor is traced off the outline of the floor below. Also, it makes the reading of the plans easier since each plan shows in the same location on the sheet, figure 8-1. The *orientation* (north direction) is thus always the same on all plan views.

ARRANGEMENT OF SHEETS IN SETS

The sheets usually are arranged in the order of their use on the job. For a small job requiring only five to ten sheets of working drawings, the following order is used.

Sheet 1 Plot Plan
Sheet 2 Foundation Plan
Sheet 3 Floor Plan
Sheet 4 Elevations
Sheet 5 Details

If there is more than one floor plan, additional sheets are used numbered 4, 5, etc. Extra sheets of elevations or details are given numbers as needed. The controlling factor is the location of the various sheets in the set. To be sure how many sheets are in the set, the reader checks the note beneath the sheet number, figure 8-2. A list of contract documents including the working drawings is found in the *Agreement*.

ARRANGEMENT OF SHEETS IN SUBGROUPS

On large projects, it is necessary to classify the drawings to a greater degree than for small projects. More-or-less standard classifications of drawings evolve because of custom and usage. The drawings for a building

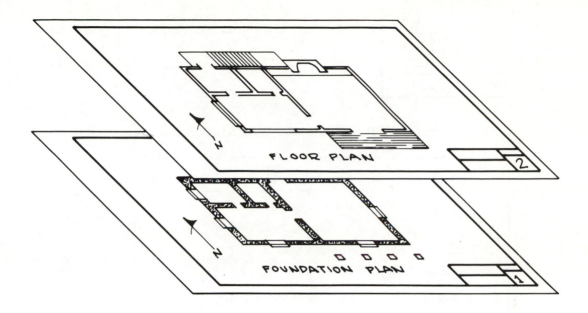

FLOOR PLAN

FOUNDATION PLAN

Fig. 8-1 One plan per sheet

RESIDENCE FOR
KERN KIMBLETON
GRAVETTE, ARK.
FLOOR PLAN
SCALE ¼" = 1'-0"

SHEET NO.

3

OF: 6

Fig. 8-2 Sheet number in set

typically include subgroups of related draw-ings. The sheets of each subgroup are labeled with a capital-letter prefix and numbered in order, figure 8-3, page 72. The number of sheets in the subgroup is noted the same as explained above for complete sets in the case of small jobs. Examples of subgroups typically used are listed in figure 8-4, page 72.

NUMBER OF SHEETS REQUIRED

Medium-sized commercial construction work requires many sheets of drawings to give the full instructions. For example, plans for a reinforced-concrete 14-story apartment tower typically have 62 sheets of working drawings. A small college student center project has 72 sheets. A university

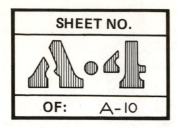 

Fig. 8-3 Subgroup capital letters

Subgroup	Notes	Prefix Letter
1. Plot Plan	Varies, sometimes in Architectural series or Topographical or Civil	A T C
2. Landscaping	Could be in Topographical or with its own symbol	L
3. Architectural	Usually next in order although might be after Structural	A
4. Kitchen	Special subgroup for institutional buildings with food services.	K
5. Structural	Includes foundation plan	S
6. Plumbing	Could be numbered with Mechanical subgroup	P
7. Mechanical	Usually covers heating and air conditioning but could include Plumbing.	M
8. Electrical	Usually last in the set. Sometimes a mechanical site plan or combination sheet labeled with two or three letters.	E ME PME

Fig. 8-4 Example of subgroups typically used

dining commons design requires 100 sheets. These plans would be too bulky if taken altogether, so they are bound in sets containing only one or two subgroups. The architectural subgroups are usually the largest containing 24 to 32 sheets for these buildings.

The size and complexity of the building determine the number of drawings. Whether there are 10 sheets or 100 sheets, however, they must be taken as a whole and understood if the building is to be built. This seems like an impossible task for a building with 100 sheets. However, each sheet can only explain a small part of the whole. Each sheet is drawn using the drafting conventions and symbols discussed in Units 1 through 6.

By careful study of each sheet related to the phase of construction being undertaken, the builder can understand just what is required. When there is uncertainty, help can be sought from the architect. At times, the architect will supply extra details.

Thus, it becomes clear to the student that commercial work involves a number of special construction methods and systems. The way the different systems of a building relate to each other is of prime importance. Units 10 through 32 give basic facts about many of the special methods and systems to aid the student in the study of building plans. Other sources of technical construction data are given in the appendix.

SUMMARY

- Working drawings is another name for blueprints.
- Architects prepare working drawings for buildings.
- Estimates and bids are based upon the working drawings and the specifications.
- Working drawings are part of the contract.
- Working drawings are used during the construction process to guide the work.
- One floor plan per sheet is a good practice and typical of working drawings.
- All plan views have the same orientation.
- The sheets of drawings usually are arranged in the order of their use on the job.
- Each sheet is numbered and the total number of sheets in the set is also indicated.
- Large building plans may have 100 or more sheets of drawings.
- Plans for large commercial buildings are divided into subgroups with special capital-letter prefixes along with the sheet numbers.
- All drawings in a set are important and must be taken as a whole to build the building.
- Architects sometimes furnish additional detail drawings during construction to further explain the work.
- The more one knows about construction methods and systems, the easier it is to read and understand blueprints for commercial construction.

REVIEW QUESTIONS

1. What is another name for working drawings?

2. Give two uses made of working drawings by the builder.

3. Which bidder usually gets the contract to build a building?
   a. the next to lowest bidder   c. the lowest bidder
   b. the bidder closest to the   d. the highest bidder
       average of the bids

4. What is the most time consuming part of the architect's work?
   a. designing a building   c. writing specifications
   b. conference with owners   d. making working drawings

5. What scale is usually used to draw plans and elevations of commercial buildings?
   a. 1/8″ = 1′-0″   c. 1/4″ = 1′-0″
   b. 3/16″ = 1′-0″   d. 1″ = 1′-0″

6. How many floor plans should be drawn on a sheet?
   a. 1   c. 3
   b. 2   d. all

7. Arrange the following sheets of working drawings in proper order.
   a. details
   b. plot plan
   c. elevations
   d. floor plan
   e. foundation plan

8. Sketch the part of a title box showing the sheet number and number of sheets in the set.

9. Pick the following capital letters that are used in this unit as prefixes to sheet numbers in plans for a large building.
   a. A
   b. D
   c. E
   d. X
   e. W
   f. M
   g. S
   h. B

10. About how many sheets of drawings does a medium-sized commercial project usually have?
    a. 200
    b. 10
    c. 25
    d. 75

11. With whom should the builder consult if a portion of a working drawing cannot be understood?

12. What additional knowledge will help the blueprint reader understand working drawings for commercial construction besides the drafting conventions and specifications studied thus far?

Unit 9
Shop Drawings

OBJECTIVES

After studying this unit, the student should be able to:

- explain what is meant by the term "shop drawings" as required in construction contracts.

- discuss shop drawing approvals giving the parts played by the contractor and the architect.

- list at least four construction items for which shop drawings are usually required.

IMPORTANCE OF SHOP DRAWINGS

Shop drawings are another type of drawing required for the construction of a modern building. It is not unheard of for the shop drawings to have as many sheets as the working drawings, sometimes more. The working drawings prepared by the architect are enough for job pricing and basic construction use. They are not suitable, however, for the production of many required items produced off the work site. Production of the job materials and equipment often requires the contract drawings to be expanded by detailed shop drawings.

Shop drawings include such fabrication, erection, and setting drawings as may be necessary to build the building. They may also include manufacturer's standard drawings or catalog cuts, performance charts, brochures and other data which illustrate the items. Samples are also required for certain materials.

PREPARATION AND APPROVAL OF SHOP DRAWINGS

These drawings usually are prepared after the contract is awarded and construction is underway. They are prepared by the vendors or fabricators of the equipment and materials and sent to the contractor for approval. The contractor checks them and sends them to the architect for approval. Only then are the materials and equipment prepared for the job.

A sufficient number of copies of each drawing must be provided so that there are enough for all parties. When the drawings are approved by the contractor and the architect, a note to that effect is signed. Approved copies of the drawings are given to the contractor who sends a copy to the vendor. Typical approved notes are shown in figure 9-1. Working from a shop drawing that does not have an approval note is a bad practice.

It is important to note that only a qualified approval is given to shop drawings by architects. Such approval relates only to design and general compliance to the drawings and specs. It is the contractor who is expected to check for quantities, dimensions, and how the items relate to other work. Field measurements such as those required for millwork and cabinets are also made by the contractor.

REQUIREMENT OF SHOP DRAWINGS

Shop drawings are required for almost every product that is fabricated away from the building site. The General Conditions (introduced in Unit 7) give the procedure for

APPROVED _____

REMARKS _____

BOWERS AND ROYCE ARCHITECTS

BY _____ DATE _____

THIS DRAWING CHECKED FOR DESIGN ONLY AND APPROVALS ARE SUBJECT TO COMPLIANCE WITH THE DRAWINGS AND SPECIFICATIONS. CONTRACTOR SHALL CHECK, VERIFY, AND SHALL BE HELD RESPONSIBLE FOR ALL MEASUREMENTS.

APPROVED
AS SUBMITTED ☐
AS NOTED ☐

AS TO CONFORMITY TO DESIGN.
CONTRACTOR TO CHECK, VERIFY
ALL QUANTITIES AND SIZES.

BY _____ DATE _____

Fig. 9-1 Approval notes on shop drawings

TYPICAL ITEMS REQUIRING SHOP DRAWINGS

Concrete Reinforcement	Structural Glazed Tile	Elevators
Structural Steel	Toilet Partitions	Fire Sprinkler Systems
Miscellaneous Metals	Suspended Ceiling Systems	Air Conditioning Equipment
Cabinetwork and Millwork	Kitchen Equipment	Electrical Power Equipment

SHOP DRAWING SPECIFICATION NOTE

Within 30 days after award of Contract and before any structural steel is fabricated or delivered to the job site, submit Shop Drawings to the Architect for approval in accordance with the General Conditions of these specifications.

Fig. 9-2

submitting shop drawings. The Specifications state which materials and equipment must be described further with shop drawings. Typical items in building construction are listed with a sample specification statement in figure 9-2.

OTHER TYPES OF SUBMITTALS

Many other items are identified by catalog references, pictures, and performance data that are already prepared. They are keyed to the architect's plans and specifications and submitted for approval. Even for an item such as finish hardware, a material's list must be submitted and approved. Every door and every piece of hardware must be listed, giving manufacturer's name and catalog number for each. This list is called a *hardware schedule* and when approved it must be

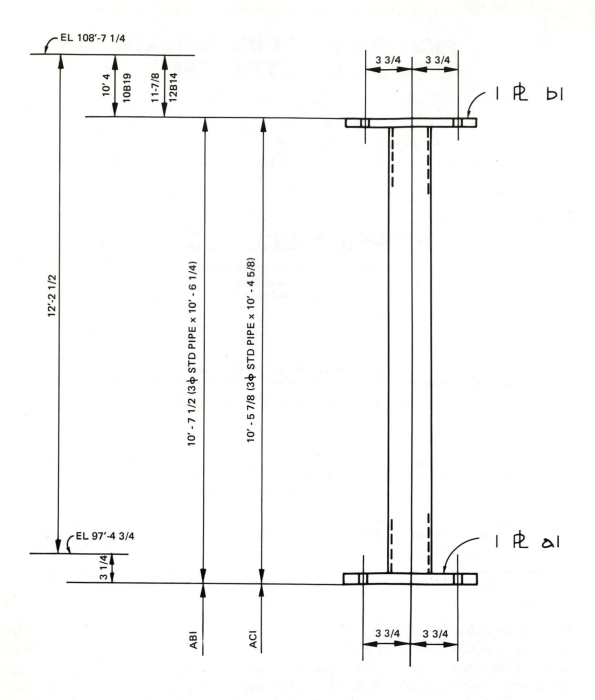

ONE COLUMN THUS AS NOTED AB1

ONE do do do do AC1

Fig. 9-3 Structural steel shop drawing

HARDWARE SCHEDULE FOR SCIENCE BUILDING

These catalog numbers are taken from catalogs of the following companies:

Butts	Stanley Works, New Britain, Conn.
Locks	Schlage Lock Co., San Francisco, Calif.
Door Stops	Glynn-Johnson Corp., Chicago, Ill.

Item 5

1 Door Hall 116 to Storage Hall 117 LHRB
3-0 x 6-8 x 1-3/4 Wd. Dr. HM Frame

1 1/2   Pr. Butts 174A5 4 1/2 x 4 1/2 1/2 MS
1   Set Locks A52PD GMK Novo x 10 STMS SM1
1   Door Stop FB 13XS US10

Item 6

1 Door Hall 116 to Elevator Shaft RHRB
1 Door Hall 115 to Elevator Shaft RHRB
2-6 x 6-8 x 1-3/8 Metal Drs. HM Frames

No Butts included
2   Set Locks A71PD GMK Novo x 10 TMS SM1
2   Door Stops FB13XS US10

CATALOG SUBMITTAL

SOLID NECK 30° ANGLE PENDANTS

RECOMMENDED: for difficult locations where there are overhead obstructions or where overhead lighting is impractical. For extra illumination of vertical surfaces.

S8N	100	Medium	8	10-7/8	10	29	8N	S-1
S10N	150	Medium	10	11-1/2	10	35	10N	S-1
S12N	200	Medium	12	13-7/16	10	42	12N	S-1
S14N	300-500	Mogul	14	15-3/16	5	24	14N	S-2
S16N	750-1500	Mogul	16	19-3/4	5	29	16N	S-2

SOLID NECK SHALLOW DOME PENDANTS

RECOMMENDED: for general illumination where lamp shielding is not important and wide distribution is desired.

S12D	100	Medium	12	7-1/8	10	32	12D	S-1
S14D	150	Medium	14	7-13/16	10	38	14D	S-1
S16D	200	Medium	16	8-13/16	5	23	16D	S-1
S18D	300-500	Mogul	18	10-13/16	5	30	18D	S-2

SOLID NECK ELLIPTICAL ANGLE PENDANTS

RECOMMENDED: for side illumination with wide distribution on both vertical and horizontal surfaces.

S8E	100	Medium	8	10-7/8	10	29	8E	S-1
S10E	150	Medium	10	11-1/2	10	35	10E	S-1
S12E	200	Medium	12	13-7/16	10	42	12E	S-1
S14E	300-500	Mogul	14	15-3/16	5	24	14E	S-2
S16E	750-1500	Mogul	16	19-3/4	5	29	16E	S-2

Lamps not furnished.

Fig. 9-4 Example of a materials list submittal

followed exactly. The carpenter who installs this hardware must read the plans for door locations and the approved hardware schedule for the exact materials to use.

Erection and setting plans are most important to the builders at the job site. The fabrication plans are included but have their best use in the manufacturer's plant. Examples of drawings of various types and other forms of submittals included in the broad scope of shop drawings are given in figures 9-3, page 78, and figure 9-4, page 79.

The drafting techniques used to prepare shop drawings are much the same as those used for the architectural plans. Most companies have some special methods or symbols, but most shop drawings are quite easy to read. Shop drawings are legal parts of the contract when approved and must be followed.

Shop drawings result in sizable costs to the manufacturer. These costs are passed on to the building owner in the form of higher prices but they are an invaluable contribution to the building process.

SUMMARY

- Shop drawings are required for many products fabricated away from the construction site.

- Shop drawings include fabrication, erection, and setting drawings.

- Shop drawing submittals may include catalog references, pictures, and performance data.

- Shop drawings are usually prepared by the vendor or fabricator after the contract is awarded.

- The contractor checks shop drawings and sends them to the architect for approval.

- The architect's approval is limited to matters of design and general compliance with the drawings and specifications.

- The specifications state which materials and equipment must be described further with shop drawings.

- The hardware submittal lists every door and piece of hardware required in the building.

- Shop drawings, although somewhat different from working drawings, are easy enough to read by the experienced blueprint reader.

- Shop drawings are legal parts of the construction contract although approved after the contract is signed.

REVIEW QUESTIONS

1. Which of the following describes a main purpose for shop drawings?
   a. job pricing
   b. basic construction
   c. construction scheduling
   d. off-site fabrication

2. The number of sheets of shop drawings necessary for commercial construction often is:
   a. 10 or less
   b. half as many as in the working drawing set
   c. as many as in the working drawing set, or more

3. Shop drawings are prepared by:
   a. the architect   c. the contractor
   b. the vendor   d. the owner

4. When are shop drawings prepared?
   a. during the bidding period
   b. after the contract is awarded
   c. at the same time as the working drawings

5. Who must approve shop drawings to make them part of the contract?

6. List four construction items for which shop drawings are usually required.

7. How are shop drawings marked to show that they are approved?

8. Shop drawing approvals are qualified, the architect being responsible for some items and the contractor for others. Which one is responsible for:
   a. dimensions   c. design
   b. quantities   d. field measurements

9. How does the contractor know which materials and equipment require shop drawings?

10. Shop drawings consist of fabrication, erection, and setting drawings. Which are most important to the builder at the job site?

11. What are some other ways to submit detailed equipment or material data than by drawings?

12. Who pays for the sizable costs of preparing shop drawings?

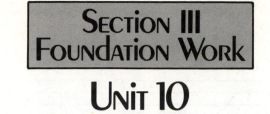

SECTION III
Foundation Work

Unit 10
Grades and Elevations

OBJECTIVES

After studying this unit, the student should be able to:

- identify a plot plan, topographic map, contours, bench mark, and grades.

- give symbols for existing contours, new contours, and spot elevations.

- calculate point elevations when given the % of grade, distance, and starting elevation.

- identify three ways to indicate the slope of construction elements.

PLOT PLANS AND TOPOGRAPHIC MAPS

Grading and layout plans show the location of the building and the site work required. These plans are also called *plot plans* and are usually the first sheets in a set of working drawings. They are developed by the architect or engineer directly from a topographic map of the site.

A *topographic map* shows the site's physical features such as trees, hills and valleys, property lines, and true north. Trees are indicated by symbols and notes, figure 10-1.

CONTOURS

Hills and valleys (the lay of the land) are indicated by contour lines. A *contour line*

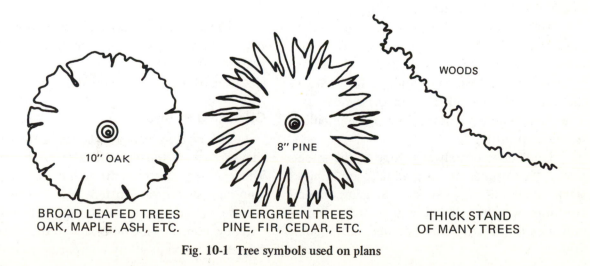

BROAD LEAFED TREES
OAK, MAPLE, ASH, ETC.

EVERGREEN TREES
PINE, FIR, CEDAR, ETC.

THICK STAND
OF MANY TREES

10" OAK

8" PINE

WOODS

Fig. 10-1 Tree symbols used on plans

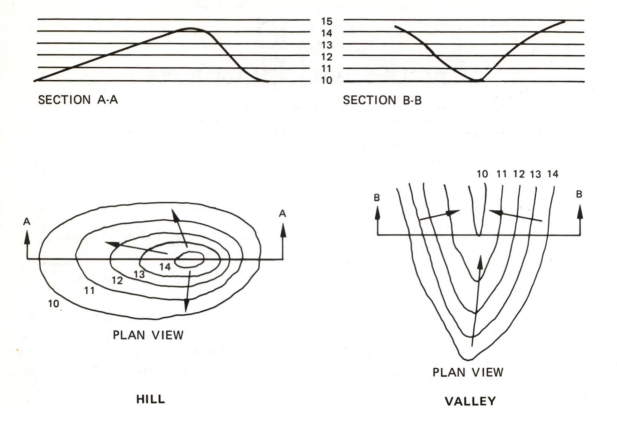

Fig. 10-2 Contour lines showing hills and valleys

is a line of constant height or elevation throughout its entire length. The pattern formed by contours of different heights indicates the shape of the land surface, figure 10-2.

The arrows on the plan views in figure 10-2 show the direction of slope. They always run from contours of higher elevation to contours of lower elevation. They show the way water would flow on these surfaces. The flow is always normal or perpendicular to the contour lines.

Land does not always slope at the same rate. Sometimes it is gradual, sometimes steep, sometimes almost flat. Spacing of the contours, whether close together or far apart, indicates the relative slope of the land, figure 10-3.

A *contour's elevation* is its height above sea level or some other reference plane. Each contour's elevation is labeled on the drawing.

The *contour interval* is the vertical distance between adjacent contours. This interval is apparent when comparing the elevation values of the contours, figure 10-4.

GRADING PLANS

The *grading plan* (plot plan) shows how the surface of the ground is to be changed. The existing contours from the topographic map are shown as dotted lines. The new contours are drawn as solid lines. Both existing and new contours are labeled with their respective elevation values. By comparing elevations

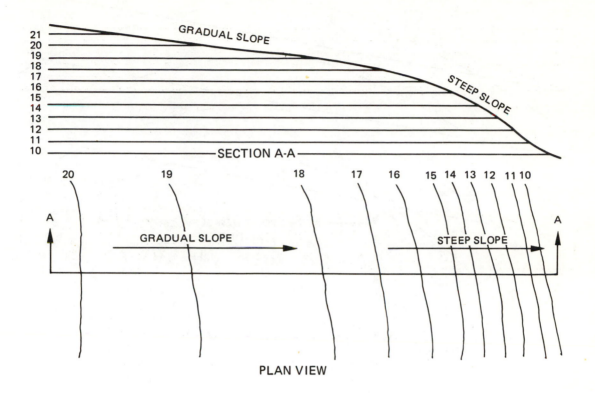

Fig. 10-3 Change in slope

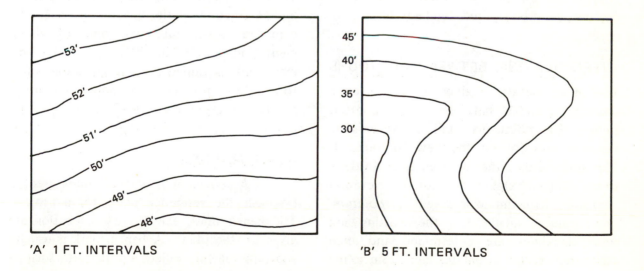

'A' 1 FT. INTERVALS

'B' 5 FT. INTERVALS

Fig. 10-4 Contour intervals

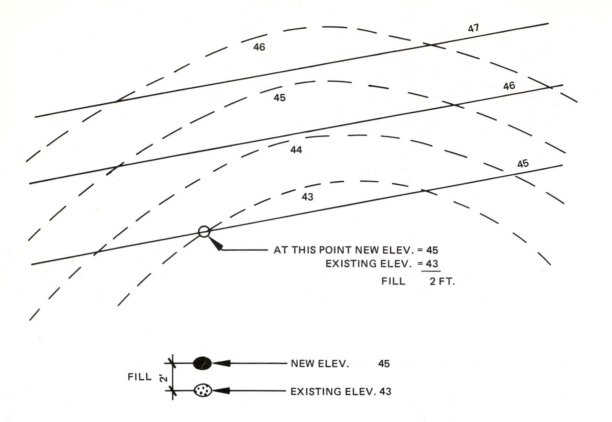

Fig. 10-5

of existing and new contours at a point, the amount of fill is determined, figure 10-5. However, if the new contour is lower than the existing one, a cut is required, figure 10-6.

INTERPOLATING BETWEEN CONTOURS

Many elevation values for points on a grading plan fall between regular contour lines. The point may fall on an existing contour but not on the new contour. It may not fall on either contour line. In these cases, it is necessary to estimate elevation values. This estimate is called *interpolation*. By separate calculations, the existing and new elevations are determined and then compared in the usual manner. Since the contour spacing is to scale, the distance between any two contours can be measured.

The distance from the lower contour line to the unknown point is also measured. Then, by multiplying the contour interval by the fraction of the distance between the contours to the unknown point, the increment (increase) of height is obtained. This increment is added to the elevation value for the lower contour to obtain the interpolated elevation, figure 10-7.

BENCH MARKS

A *bench mark* is a fixed point used to establish the reference plane for a project. The bench mark elevation and location are given on the plan. A bench mark is usually some permanent object on the site such as the top of a manhole in the street near the project. Its elevation may be arbitrarily

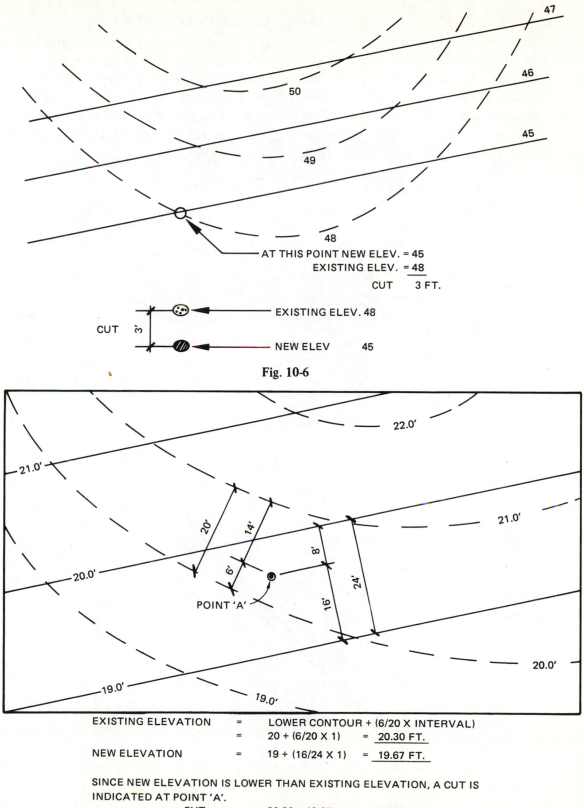

AT THIS POINT NEW ELEV. = 45
EXISTING ELEV. <u>48</u>
CUT   3 FT.

CUT   3'   EXISTING ELEV. 48
            NEW ELEV   45

Fig. 10-6

EXISTING ELEVATION	=	LOWER CONTOUR + (6/20 X INTERVAL)	
	=	20 + (6/20 X 1)	= <u>20.30 FT.</u>
NEW ELEVATION	=	19 + (16/24 X 1)	= <u>19.67 FT.</u>

SINCE NEW ELEVATION IS LOWER THAN EXISTING ELEVATION, A CUT IS INDICATED AT POINT 'A'.

CUT   =   20.30 – 19.67   = <u>.63 FT.</u>

Fig. 10-7 Contour interpolation

assigned, such as 50.0 ft. or 100.0 ft. The bench mark elevation may be based upon sea level. Sometimes when sea level elevations are used, the first two digits are dropped for simplicity. Thus, 1235.0 ft. is written 35 ft. and other elevations are written similarly. When this abbreviated method is used, a note is given to that effect on the plans, figure 10-8.

GRADE CALCULATIONS

Most grading plans for buildings indicate contours at one-foot intervals. The distance between contours is drawn to the same scale as all the other features on the plan. The true slope can be shown by drawing a section at right angles to the contours, figure 10-9.

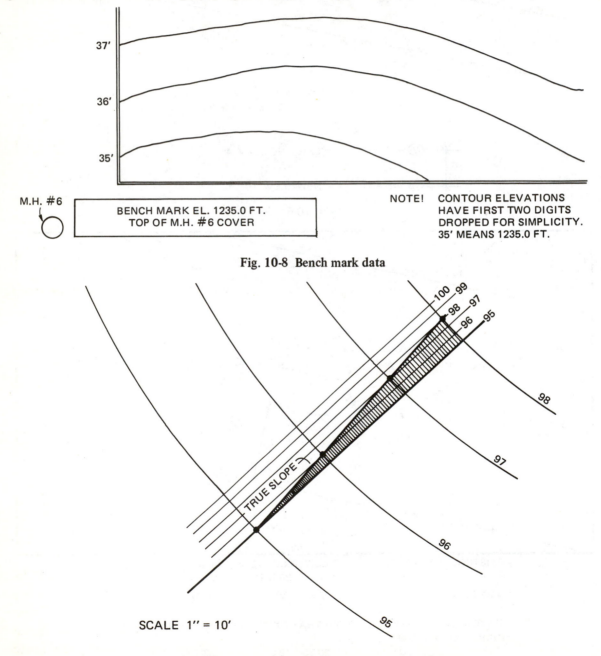

Fig. 10-8 Bench mark data

Fig. 10-9 One method for calculating true slope

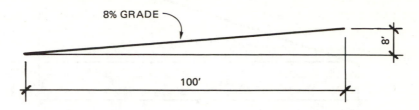

Fig. 10-10 Mathematical ratio method for calculating true slope

Although this procedure is quite accurate, it is too time consuming to use routinely. A mathematical ratio method used by the engineer is more useful. In this method, the slope is defined in terms of % (percent) of grade. Simply stated, an 8% grade means a rise or fall of 8 ft. per 100 ft., figure 10-10. When grade calculations are made, percent values are changed to the decimal form by dividing them by 100, for example 8% = .08.

The following formulas are used to calculate grade problems. When any two values are known, the third value is found by using the proper formula. Their use is illustrated by examples and problems which follow.

EXAMPLES AND PROBLEMS

Formula 1. $grade = \dfrac{rise\ (or\ fall)}{distance}$

Formula 2. rise (or fall) = grade x distance

Formula 3. $distance = \dfrac{rise\ (or\ fall)}{grade}$

Problem A The plot plan indicates the high end of a 120 ft. straight curb to be at elevation 75.8 ft. The low end is at elevation 74.0 ft. What fall does each 16 ft. section of the curb have?

Solution: Step 1 Calculate the grade (Formula 1)

$grade = \dfrac{fall}{distance} = \dfrac{(75.8 - 74.0)}{120} = 0.15\ (1.5\%)$

Step 2 Calculate the fall for a 16 ft. interval (Formula 2)

fall = grade x distance = .015 x 16 = *.24 ft. ans.*

Problem B Determine the elevation of a sewer pipe where it passes through the foundation wall. The street sewer is at elevation 42.5 ft. The grade of the pipe is .01 (1%) and the distance is 150 ft.

Solution: Step 1 Calculate the rise from street to foundation (Formula 2)

rise = grade x distance = .01 x 150 ft. = 1.5 ft.

Step 2 Add rise to street elevation 1.5 + 42.5 = *44.0 ft. ans.*

Problem C How long is a sewer pipe if it falls 6 ft. and slopes at 2% grade?

Solution: Step 1 Calculate the length of the sewer pipe (Formula 3)

$distance = \dfrac{fall}{\%\ grade} = \dfrac{6\ ft.}{.02} = 300\ ft.$

PIPE GRADES

The grade of a surface or pipe may be given as a ratio (1/8 in. per ft.). To relate this data to elevations given on the plans, the data must be changed to the decimal form. Since 1/8 in. is approximately .01 ft., the grade is easily calculated by using Formula 1.

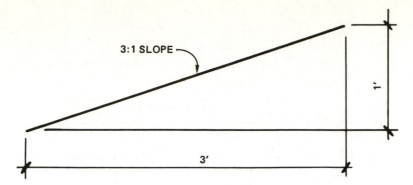

Fig. 10-11

$$grade = \frac{\text{rise (or fall)}}{\text{distance}} = \frac{.01 \text{ ft.}}{1 \text{ ft.}} = .01 \ (1\%)$$

In like manner 1/4 in. per ft. = 2/8 = .02 (2%)

Also 1/2 in. per ft. = 4/8 = .04 (4%)

OTHER SLOPE CONVENTIONS

Slope is sometimes given as a ratio of the horizontal to the vertical distances such as 3 : 1, figure 10-11. In this case, the ground falls 1 ft. for each 3 ft. of horizontal distance.

The slope of concrete slabs is indicated by an arrow. The total fall across the slab is indicated by a number placed above the arrow, figure 10-12. The fall from the normal slab elevation is indicated thus (–3/4″), figure 10-13. When floor drains are involved, lines are drawn from the corners of the room to the drain.

SPOT ELEVATIONS

Spot elevations as shown in figure 10-14 are also used to establish the heights of elements. They give precise elevations at definite points on drives, curbs, walks, lawns, etc. They are also used to indicate slopes of porches, terraces, patios, and roofs.

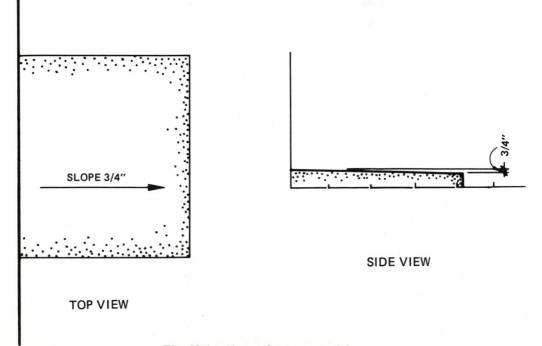

Fig. 10-12 Slope of a concrete slab

A good set of plans has many parts labeled with grade and elevation values. Without these values clearly stated to control the work, few modern buildings could be constructed. Knowing the precise meaning of the symbols used to convey this data is of great importance to the builders of today.

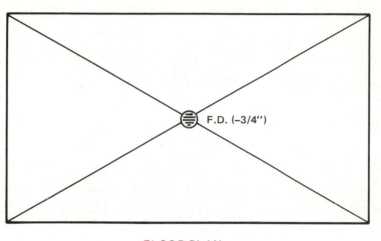

FLOOR PLAN

Fig. 10-13 Slope of concrete slab when floor drain is involved

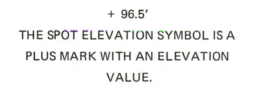

+ 96.5′

THE SPOT ELEVATION SYMBOL IS A

PLUS MARK WITH AN ELEVATION

VALUE.

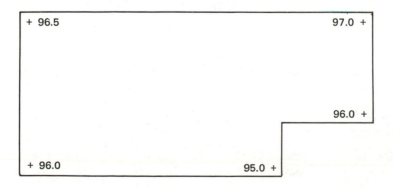

Fig. 10-14 Spot elevations

SUMMARY

- Plot Plans show the location of a building on its site.

- A topographic map showing the physical features of an existing site is used by the architect or engineer to develop the Plot Plan.

- Hills and valleys are indicated by contour lines on topographic maps.

- All points along a contour line are of the same height or elevation.

- Water flow on a surface is always normal (perpendicular) to the contours.

- Spacing of contours, whether close together or far apart, indicates the relative shape of the land.

- A contour's elevation is its height above sea level or some other reference plane.

- The contour interval is the vertical distance between adjacent contours.

- When the surface of the ground is to be changed, new contours are shown as solid lines and the existing contours that are changed are shown as dotted lines.

- When the new contour is higher than the existing one, fill is indicated.

- If the new contour is lower than the existing one, a cut is required.

- Elevation values of points that fall between contour lines are found by interpolation.

- A bench mark is a fixed point used to establish the reference plane for a project.

- Most grading plans for buildings indicate contours at one-foot intervals.

- The slope of the ground can be given in terms of % (percent) grade.

- Grade is calculated by dividing the difference in elevation of two points (rise or fall) by the distance between the points.

- The slope of a pipe may be given as a ratio (1/8 in. per foot).

- Slope is sometimes given as a ratio of the horizontal distance to the vertical distance of one such as 2 : 1 or 3 : 1.

- Sloping concrete slabs can be indicated by slope arrows with the amount of fall noted.

- Spot elevations are used to indicate precise elevations of points on a surface or element.

- Grade and elevation values are used to control the work in modern building plans and are essential to their construction.

REVIEW QUESTIONS

Questions 1—5 refer to Topographic map, figure 10-A.

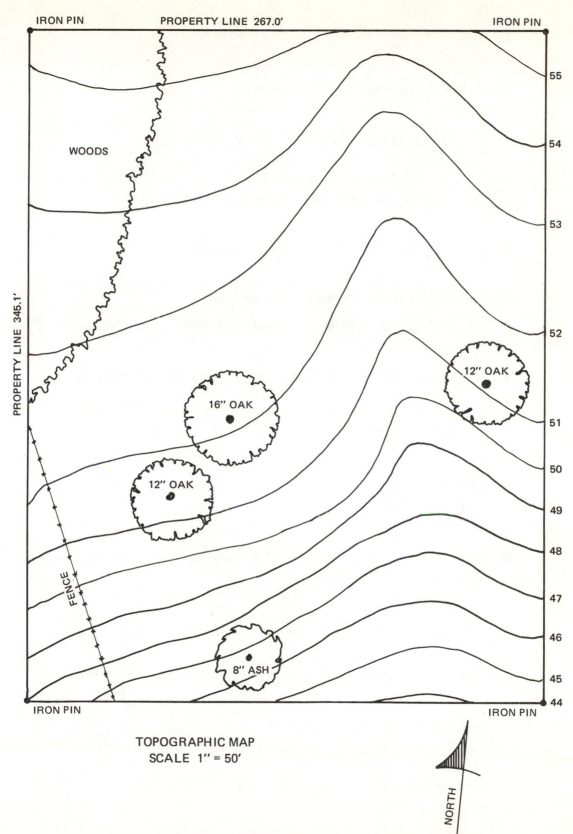

Fig. 10-A

1. What is the dimension of the north property line?

2. What is the contour interval on this map?

3. In what general direction does the land slope downhill?

4. How far in feet is it between the two 12 in. oak trees?

5. What is the maximum slope from north to south?

Questions 6–10 refer to Grading plan, figure 10-B.

6. What is the reference plane elevation for this plan?

7. How much fill is required at the northwest corner of the property?

8. Will more water drain out of the driveway at A or at B?

9. What is the interpolated elevation of the original ground at the north-west corner of the building?

10. What is the elevation of the top of the new manhole if it is level with the ground?

For questions 11–15, use information and formulas concerning Grade calculations.

11. Most grading plans for buildings show contours at what intervals?

12. What is the decimal form for 1.5%?

13. How many feet per 100 ft. does an 8% grade rise?

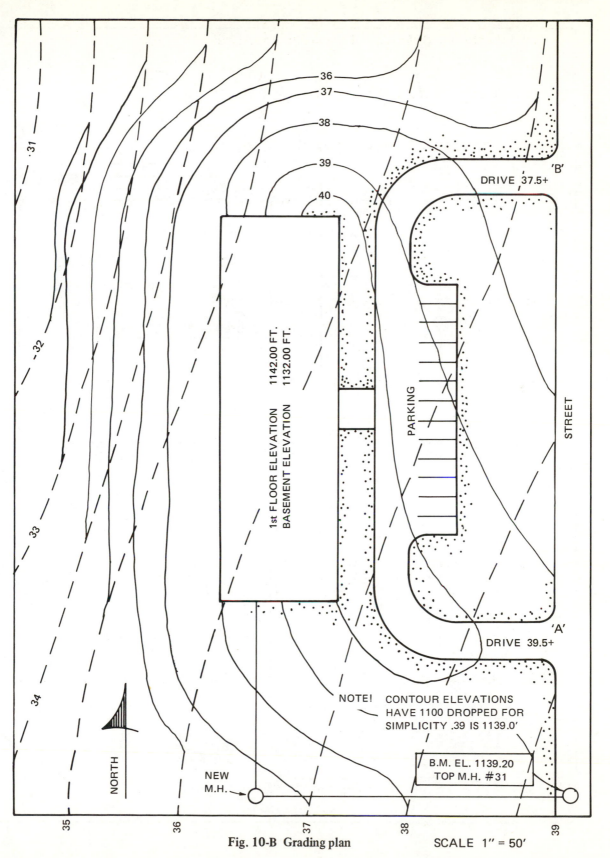

Fig. 10-B Grading plan   SCALE 1″ = 50′

Labels within figure:

36
37
38
39
40

'B'

DRIVE 37.5+

31
32
33
34

1st FLOOR ELEVATION   1142.00 FT.
BASEMENT ELEVATION   1132.00 FT.

PARKING

STREET

'A'

DRIVE 39.5+

NOTE! CONTOUR ELEVATIONS
HAVE 1100 DROPPED FOR
SIMPLICITY .39 IS 1139.0′

B.M. EL. 1139.20
TOP M.H. #31

NORTH

NEW
M.H. →

35   36   37   38   39

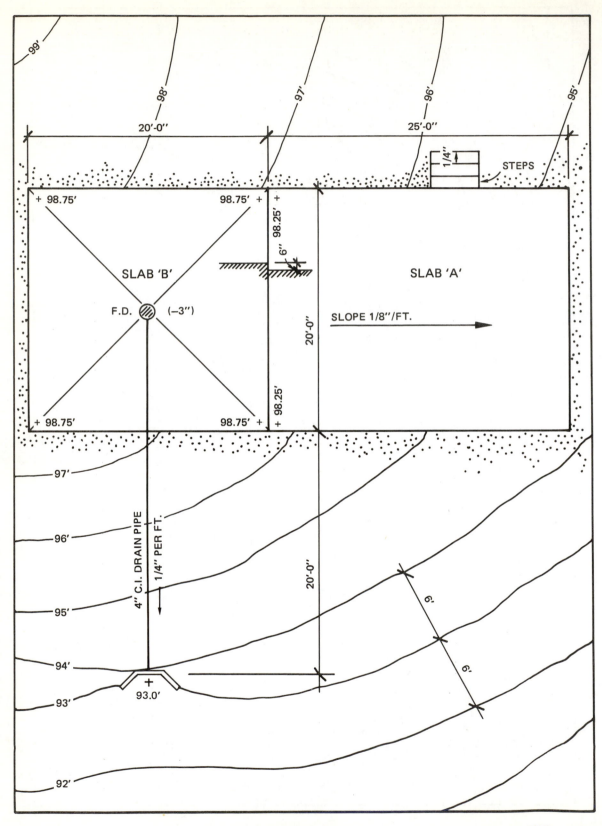

Fig. 10-C Slab Plan   SCALE 1/8″ = 1′-0″

14. Using Formula 1 calculate the grade when the rise is 2 ft. and the distance is 50 ft.

15. A point on a grading plan lies 15 ft. from a 49.0 ft. contour and 45 ft. from a 50 ft. contour. What is the elevation of the point?

For questions 16-20 refer to Slab Plan, figure 10-C, page 96.

16. What is the slope of slab A?

17. How much fall is there to the floor drain in slab B?

18. What % grade is the slope of slab A?

19. What slope does the drain pipe from the floor drain have? What % grade is this?

20. What is the elevation of the outlet?

Unit 11
Building Layout

OBJECTIVES

After studying this unit, the student should be able to:

- describe the steps required to make an accurate layout.

- give three methods used to square a layout.

- tell how columns are located and identified on structural plans.

LAYOUT DEFINED

One of the most important steps in the construction of a building is the layout. In a way this is like drawing the outline of the building, full-size, on the ground. In the case of excavations, the lines are often drawn with lime making bright white markings. These are followed easily by the machine operator as the trench or hole is dug, figure 11-1.

MEASUREMENTS

The methods used for making accurate layouts involve measuring, squaring up, and preserving the points and lines. Lines drawn on the ground with lime are too coarse for accurate layouts. Measurements are made with steel tapes and are generally accurate to the nearest 1/8 of an inch. This corresponds to the minimum dimensional unit found on

Fig. 11-1 Backhoe excavating to limed layout line

the drawings for floor plans, foundations, and similar views.

When beginning the layout, an important step in reading blueprints is to check the dimensions. Sometimes the architect includes a note on the plans such as the following:

Contractor shall be responsible for the accuracy of all dimensions and shall notify the architect promptly of any errors.

This means that the contractor must have someone check to see that strings of minor dimensions add up to equal overall dimensions. Sometimes this is a difficult task, but errors found after the layout is begun can be costly. The error may be the fault of the architect, but the contractor pays if the error is not found in time. The contractor also pays if the plans are not read properly and errors result.

Dimensions in layouts are recorded in feet and inches; therefore, the addition of mixed numbers as found on building plans needs to be practiced to perfection by the commercial builder. Calculators are not common that add feet and inches and fractions of inches unless changed to the decimal form. This is an awkward conversion and not generally done. A suggested method to

simplify the addition of mixed numbers is shown in figure 11-2. The standard method is shown for comparison. Most foreign countries use the metric system now. In years to come the metric system is expected to be adopted by the construction industry in the United States. This is a decimal system and simplifies dimensional calculations.

The dimensions used for the layout of the building are found on the plot plan, foundation plan, and basement or first floor plan. The plot plan shows the location of the building and other features of the site. The property lines are used to locate the position of the building. They are required on the plot plan so that the setbacks required by zoning laws can be checked when the building permit is issued. Some major building dimensions usually are shown on the plot plan, figure 11-3, page 100.

The best source of layout dimensions for actual construction is the foundation plan. This usually is found in the structural subgroup with prefix S. The foundation plan, as well as other structural drawings, is drawn to provide the minimum essential information for construction purposes, figure 11-4, page 101. At first glance, the structural drawings and foundation plan may seem incomplete. They show very little of the detail that the

Fig. 11-2 Addition of mixed numbers

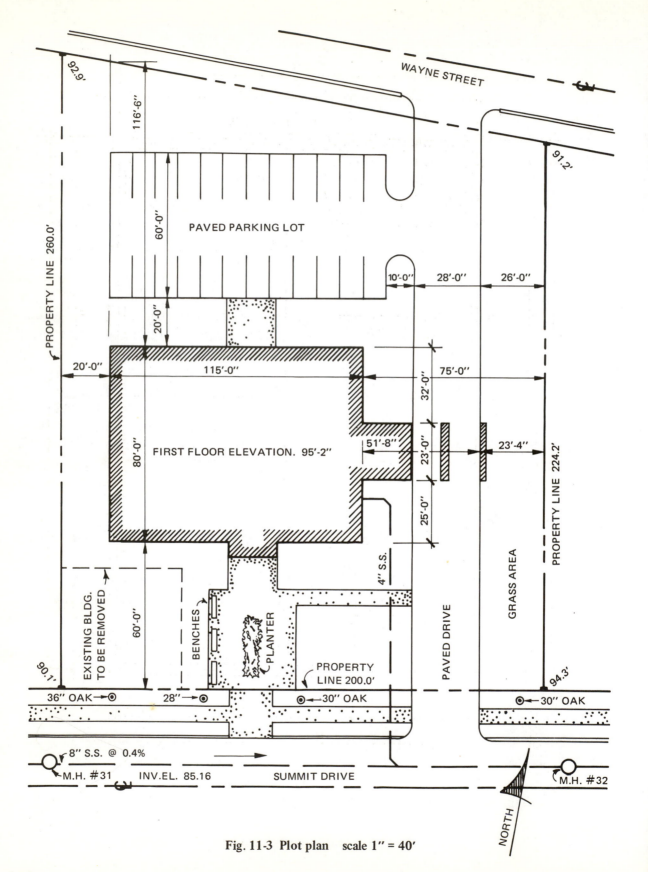

Fig. 11-3 Plot plan   scale 1″ = 40′

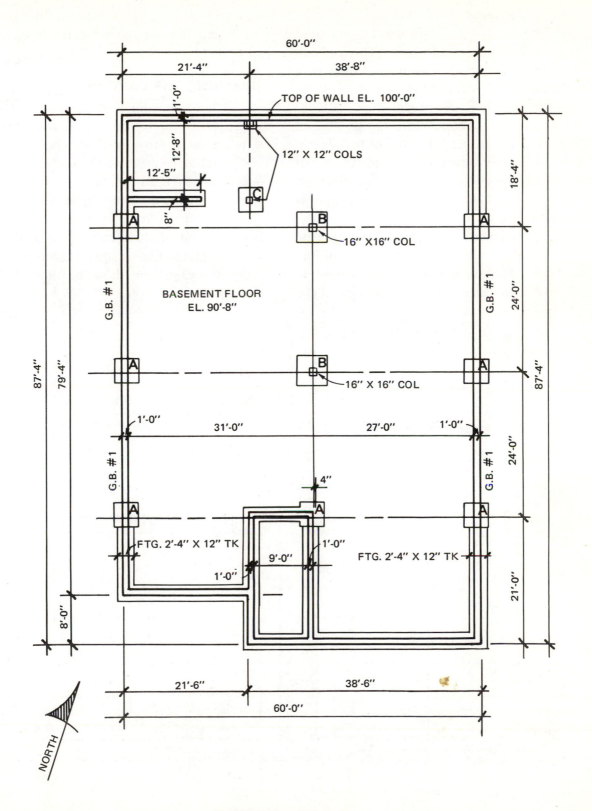

Fig. 11-4 Foundation plan

architectural sheets show. For instance, often footings are drawn using continuous solid lines instead of dotted lines. This convention is used because of the ease of drawing a continuous line compared to a dotted line. This convention is also proper if we are looking at the foundation before the backfill or floor slabs are placed. The top of foundation walls may be shown instead of the usual horizontal section view of architectural sheets. No material sectioning symbol is needed in this case.

The foundation is dimensioned to the faces of the walls, not to the edge of the footings. These major dimensions should be compared to those on the floor plan. The wall sections give more information relating the foundation to the superstructure, figure 11-5.

SQUARING THE LAYOUT

After the main part of the building is staked using the dimensions from the plans, it is necessary to check the layout. The best check is to measure the diagonals of the largest rectangle in the layout. If these are equal, the figure is square, figure 11-6.

To help make a square layout on the first try, one of three methods is used. The simplest method is to square the first corner using the 3-4-5 method as illustrated in figure 11-7.

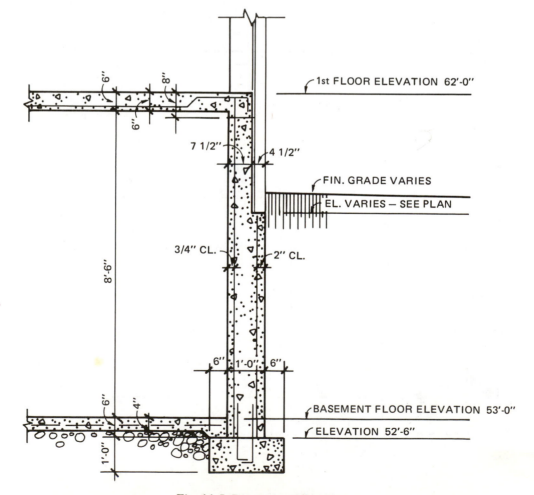

Fig. 11-5 Basement wall section

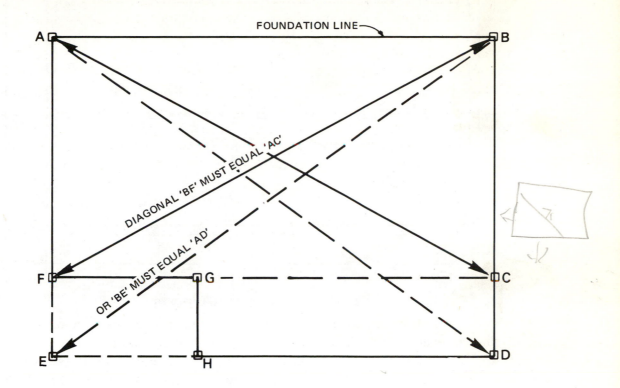

Fig. 11-6 Diagonal check of layout

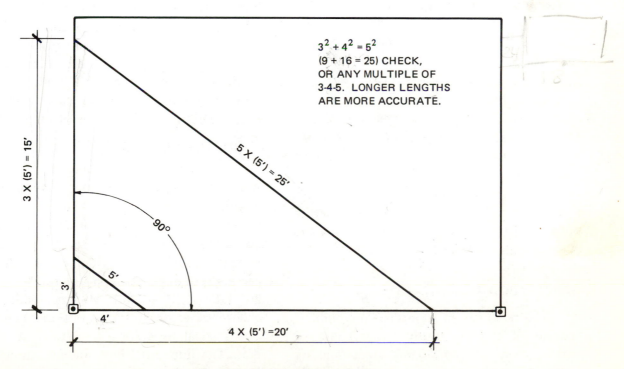

Fig. 11-7 3-4-5 squaring method

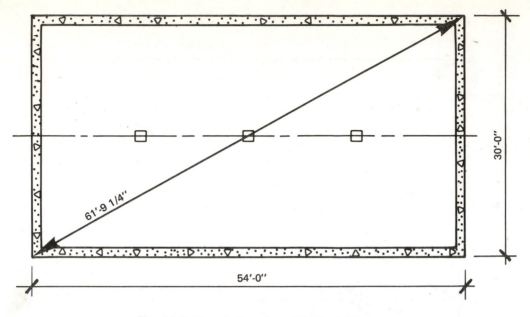

Fig. 11-8 Foundation plan with diagonal

A better method is to compute the length of the diagonal. By using two steel tapes, one for the side and one for the diagonal, the unknown corner can be located exactly. Computing the diagonal is done by squaring each side, adding the squares together and then taking the square root. Small calculators are available to do this easily. It would be helpful if the drafter would show the diagonal dimension on the drawing. In rare cases this is done, figure 11-8.

When the building is too large for these methods, the corners are squared using a transit. The instrument is set directly over one corner. The transit is indexed to the known line by sighting down the line to the other corner. Then by rotating the instrument 90°, a new line is established. Any number of points can then be located by measuring and aligning with the transit, figure 11-9.

BATTER BOARDS

The final step in laying out the building consists of erecting sturdy batter boards at each corner. *Batter boards* are temporary horizontal supports erected near the corners

of the building layout to hold lines representing the building outline. These boards are set four feet or more outside the building line, and the layout is transferred to them, figure 11-10. In this way, the layout is preserved when footings are dug. The tops of the batter boards should be set level with each other at a selected height. The batter boards usually are set at the floor-line elevation or top of the foundation. The plot plan gives the elevation of floors as well as the bench mark to which they are referenced. In any case, in the layout, a careful reading of the plans is critical.

STRUCTURAL GRID

Reinforced concrete and structural steel framed buildings require a special layout called a structural grid. The plans are drawn with a system of reference lines for columns so that all details can be referred to this grid. Most columns are located by grid lines through their centers. All dimensions on structural drawings are referenced to this grid to reduce the chance of error in laying out the building. The vertical lines are given letters, A, B, C, etc. The horizontal lines are given numbers, 1, 2, 3, etc., figure 11-11, page 106.

Every column is identified by its vertical and horizontal grid line name, such as B2, C4, D5, etc. These grid lines are labeled on the batter boards and become the master grid for the construction of the building. Architectural plans and elevations show these structural grids also and use them to control the construction.

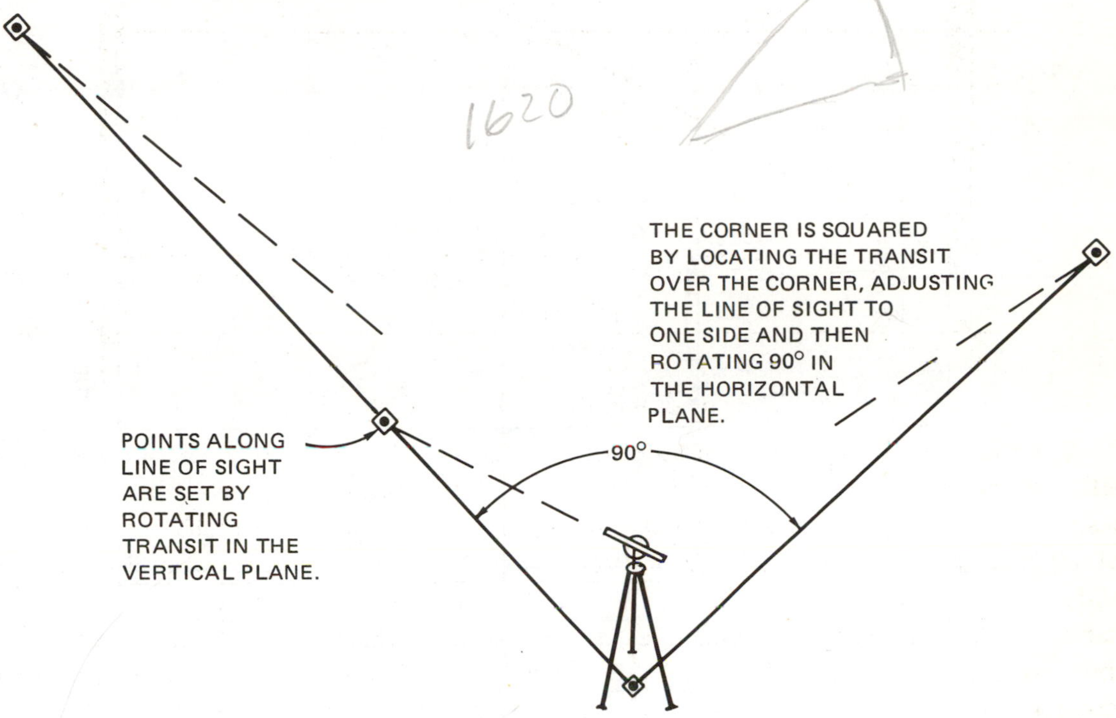

THE CORNER IS SQUARED BY LOCATING THE TRANSIT OVER THE CORNER, ADJUSTING THE LINE OF SIGHT TO ONE SIDE AND THEN ROTATING 90° IN THE HORIZONTAL PLANE.

POINTS ALONG LINE OF SIGHT ARE SET BY ROTATING TRANSIT IN THE VERTICAL PLANE.

90°

Fig. 11-9 Using the transit for layouts

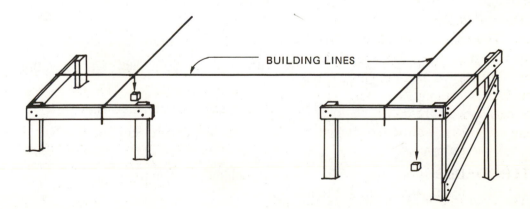

BUILDING LINES

Fig. 11-10 Batter boards

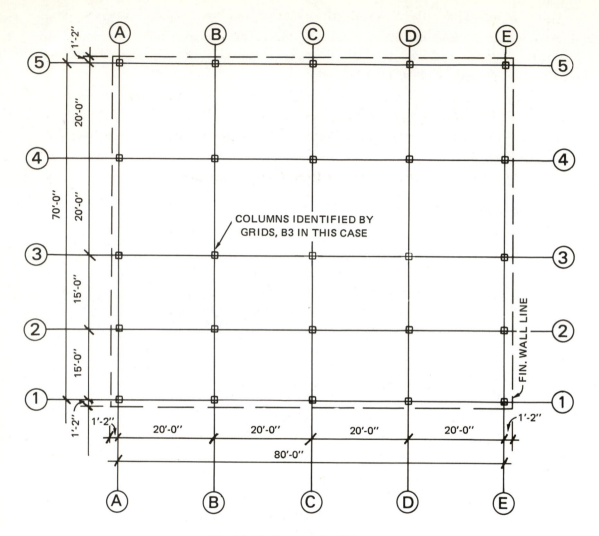

Fig. 11-11 Structural grid layout

SUMMARY

- An accurate layout is essential to the construction of any building.

- Major dimensions are made with steel tapes and accurate to within 1/8″ of the true dimension.

- Contractors are required to verify and check all dimensions.

- The addition of mixed numbers is an important skill for the builder to have.

- Dimensions for the layout of a building are found on the plot plan, foundation plan, and first floor plan.

- Structural foundation plans show minimum details and the footing outlines appear as solid lines.

- Wall sections are important sources of information when making a layout.

- Measuring the diagonals of a layout is the best check of its squareness.
- A transit is used to lay out large buildings.
- Batter boards are used to preserve the main layout during the construction of the foundation.
- Batter boards usually are set level to each other and at the floor or foundation wall elevation.
- Structural framed buildings have their columns located by means of a grid.
- Structural columns are identified by their grid point name.

REVIEW QUESTIONS

1. Rough layout lines for excavating purposes are marked by:
   a. scratching a line on the ground
   b. setting a row of small stones
   c. sprinkling lime in a line on the ground
   d. stretching a string between two stakes

2. Layouts are made with dimensions accurate to the nearest
   a. 1/2 inch
   b. 1/4 inch
   c. 1/8 inch
   d. 1/32 inch

3. Who is responsible for the accuracy of all dimensions after the construction begins?
   a. contractor
   b. architect
   c. owner
   d. insurance company

4. Add the following dimensions:

   a. 4'-3"
   5'-9"
   10'-6"
   15'-2"
   8'-4"

   b. 4'-6"
   6'-2 1/2"
   5'-8"
   7'-4"
   15'-6 1/4"

   c. 20'-0"
   13'-6"
   10'-8"
   11'-4"
   16'-0"

5. Of the following plans, which is not used in the layout of a building, foundation plan, roof plan, floor plan, basement plan?

6. Name three methods used to square a building layout.

7. A foundation wall layout is made to:
   a. the outside of the footings   c. the exterior face of the wall
   b. the center line of the wall   d. the interior face of the wall

8. Batter boards are used to preserve the layout during construction and are set with their tops:
   a. four feet above the ground
   b. level with the top of the footing
   c. level with each other at any height above the floor
   d. level with the floor line or top of foundation

9. Columns in large buildings using a structural grid layout are identified by:
   a. drawing them to scale with the proper material indication
   b. a note giving the size
   c. schedules in the specifications
   d. the grid point name

10. Name three steps involved in making accurate building layouts.

Unit 12
Footings–Wall and Column

OBJECTIVES

After studying this unit, the student should be able to:

- explain why it is important to locate footings directly under the walls and columns they support.

- compare the projections of plain concrete footings with reinforced footings.

- tell how the size of column footings is given in the plans.

FOOTING SIZE

After the main building lines are located, the footings can be marked out. *Footings,* sometimes called spread footings or pads, are enlarged areas of concrete designed to distribute loads to the ground. By spreading out, they can support more weight without sinking too much. Size of the footing is figured for the type ground and the loads to be carried. This is more important in the case of column footings than for wall footings. Wall footings are usually continuous along the wall. The wall helps spread heavy-point loads to the footing, thus averaging out the pressure, figure 12-1. Column footings, however, are most often square and sized for the column load. There is no simple way to spread the column load to some other part of the foundation. Therefore, the column footing carries the whole load and must be sized carefully, figure 12-2, page 110.

FOOTING LOCATION

Another important factor in footing layout is the need to locate footings exactly according to the plans. By this is meant, not just under the wall or column but concentric (having a common center) with the wall or column. If a column footing is not located under the center of the column, the pressure will be uneven. The footing may be the size indicated on the drawings, but the uneven

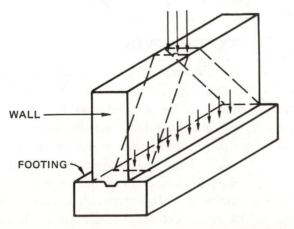

WALL

FOOTING

CORBELLING EFFECT
SPREADS POINT LOAD
ON THE FOOTING A
DISTANCE AT LEAST
EQUAL TO WALL HEIGHT.

Fig. 12-1 Wall with point load

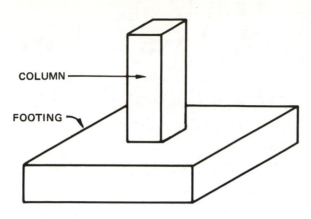

COLUMN

FOOTING

COLUMN POINT LOAD
SPREADS THROUGH THE
FOOTING BY SHEAR
AND BENDING.

Fig. 12-2 Column footing

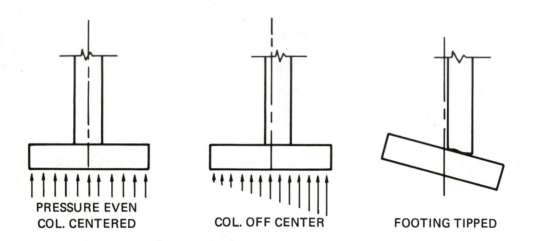

PRESSURE EVEN
COL. CENTERED

COL. OFF CENTER

FOOTING TIPPED

Fig. 12-3 Footing fails by tipping

pressure could cause the footing to tip, figure 12-3.

Careless over or undersizing of footings usually results in their being out of position. There are some footings that are off-center because of a special situation. The cantilever retaining wall is an example of a continuous wall footing being offset, figure 12-4. This is done in order to control the pressure since there is a tendency for the wall to turn over. The earth being supported pushes laterally on the wall.

In addition to the need for equal pressure under the footings is the need to have them level. The excavation must be made level before forms for footings are built. This

applies to lateral as well as lengthwise directions, figure 12-5. Wall footings may need to be stepped to fit sloping ground, figure 12-6.

FOOTING DEPTH

Footing depth, or thickness, is also designed to meet load and material requirements. Unreinforced wall footings (footings with no cross bars) are usually proportioned with a depth equal to that of the wall thickness. They are double the wall in width which makes them project one half the wall thickness. This provides a safe proportion to resist cracking under load, figure 12-7, page 112.

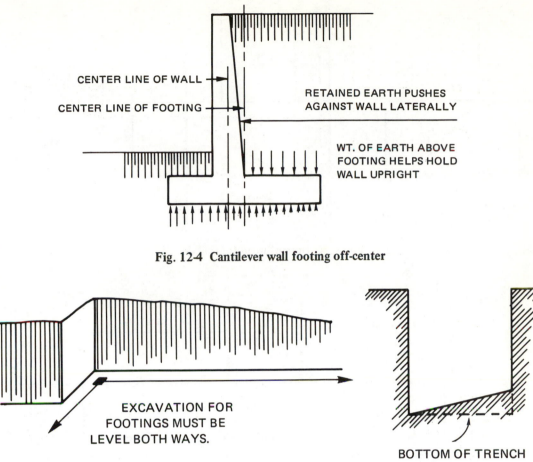

CENTER LINE OF WALL

CENTER LINE OF FOOTING

RETAINED EARTH PUSHES
AGAINST WALL LATERALLY

WT. OF EARTH ABOVE
FOOTING HELPS HOLD
WALL UPRIGHT

Fig. 12-4 Cantilever wall footing off-center

EXCAVATION FOR
FOOTINGS MUST BE
LEVEL BOTH WAYS.

BOTTOM OF TRENCH
FOR FOOTING MUST
BE CUT LEVEL.

Fig. 12-5 Footing bearings must be level

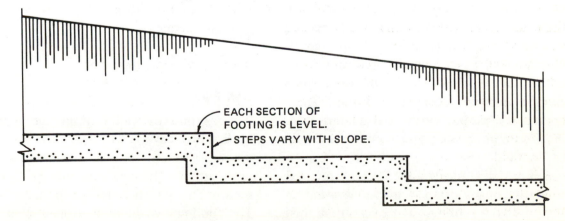

EACH SECTION OF
FOOTING IS LEVEL.
STEPS VARY WITH SLOPE.

Fig. 12-6 Stepped footing used on slopes

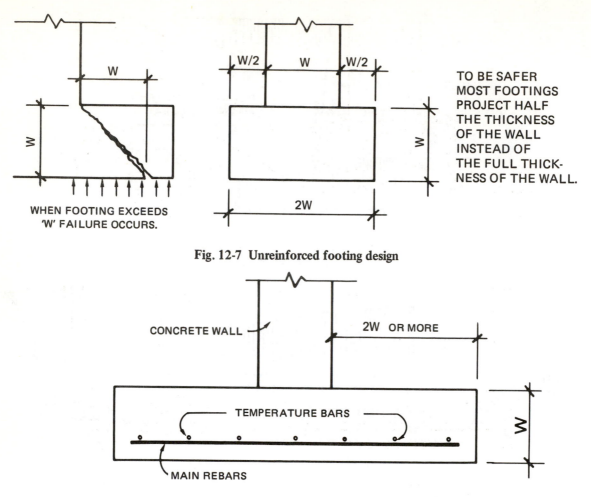

Fig. 12-7 Unreinforced footing design

WHEN FOOTING EXCEEDS 'W' FAILURE OCCURS.

TO BE SAFER MOST FOOTINGS PROJECT HALF THE THICKNESS OF THE WALL INSTEAD OF THE FULL THICKNESS OF THE WALL.

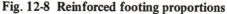

Fig. 12-8 Reinforced footing proportions

Reinforced concrete footings project much more than one half the depth of the footing. This is possible because of the effect of the steel bars placed at special locations, figure 12-8. Almost all column footings are reinforced. Continuous wall footings sometimes have two bars running lengthwise of the footing, figure 12-9. These bars do not help hold the projecting parts of the footing. These footings are proportioned as if unreinforced.

Footings should never be made less than the depth shown on the plans. They can be deeper but this adds to the cost. Structural foundation plans often give the elevation of the top of the footing. The excavation must be deep enough to provide for the proper amount of concrete. If it is dug deeper than required, it must be filled with concrete. Because of the cost of concrete, final trimming of footing holes often is done by hand to make them exact.

FOOTING CONVENTIONS

Wall Footings

Continuous wall footings are indicated in various ways. They are drawn in the proper position on the foundation plan using either light continuous lines or dotted lines. The size is noted as in figure 12-10. The footing may be dimensioned as in figure 12-11. A reference to a foundation section

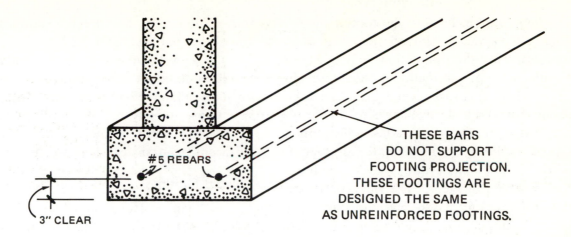

#5 REBARS

3" CLEAR

THESE BARS
DO NOT SUPPORT
FOOTING PROJECTION.
THESE FOOTINGS ARE
DESIGNED THE SAME
AS UNREINFORCED FOOTINGS.

Fig. 12-9 Typical footing reinforcement

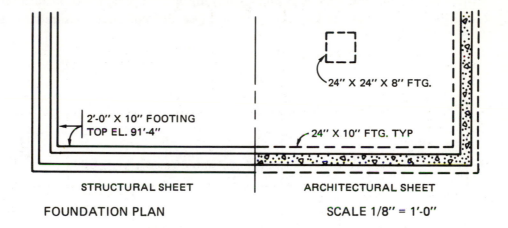

24" X 24" X 8" FTG.

2'-0" X 10" FOOTING
TOP EL. 91'-4"

24" X 10" FTG. TYP

STRUCTURAL SHEET

FOUNDATION PLAN

ARCHITECTURAL SHEET

SCALE 1/8" = 1'-0"

Fig. 12-10 Footing size by note

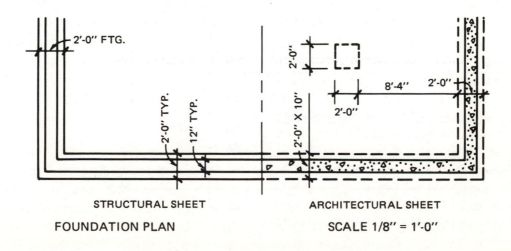

2'-0" FTG.

2'-0" TYP.

12" TYP.

2'-0" X 10"

2'-0"

8'-4"

2'-0"

2'-0"

STRUCTURAL SHEET

FOUNDATION PLAN

ARCHITECTURAL SHEET

SCALE 1/8" = 1'-0"

Fig. 12-11 Footing size dimensioned

also helps the reader find the footing dimensions. When the footing continues the same size all around the foundation, only one note or dimension is required. It should be obvious when a change occurs, and the note beside it should indicate a new size.

Steps in the continuous wall footing should be shown on the plan. If the situation is complicated, an elevation view of the foundation shows the steps. The elevation of the top of the footing may be given, or sometimes the bottom elevation is indicated. In other cases, only the elevation of the basement floor is given, and the footing top elevation must be calculated. Again, the section is an excellent source for these details. In any case,

information must be found to establish the top of the footings, figure 12-12.

If a wall does not have continuous spread footings, it may have spaced footings or piers, figure 12-13. In rare cases, light structures do not require footings of any kind. The width of the foundation wall is sufficient to transfer the load to the soil.

Column Footings

Column footings are located by the column center lines. All these footings should be centered under the column they support unless specifically noted. The outline of the footing is drawn in position using light solid lines or dotted lines. Rarely is a column

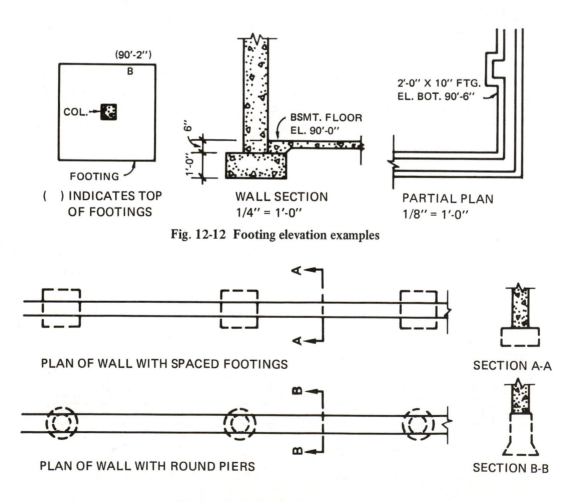

Fig. 12-12 Footing elevation examples

Fig. 12-13 Spaced footings and piers

MARK	FOOTING SIZE	REINFORCING	PEDESTAL
A	4' - 8" x 4' - 8" x 1' - 0"	9 - #4 x 4' - 2" E.W.	NONE
B	8' - 0" x 8' - 0" x 1' - 4"	19 - #5 x 7' - 6" E.W.	4' - 6" x 4' - 6" x 10"
C	6' - 4" x 6' - 4" x 1' - 3"	12 - #5 x 5' - 10" E.W.	NONE
D	7' - 0" x 7' - 0" x 1' - 3"	17 - #5 x 6' - 6" E.W.	NONE
E	5' - 6" x 5' - 6" x 1' - 1"	12 - #4 x 5' - 0" E.W.	NONE
F	2' - 4" x 2' - 4" x 1' - 0"	4 - #5 DOWELS	NONE

Fig. 12-14 Sample footing schedule

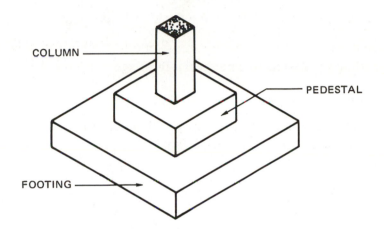

Fig. 12-15 Column footing with pedestal

footing dimensioned. A note may be used, but the best method is a footing schedule, figure 12-14.

Usually there are several column footings of the same size. The schedule simplifies the drafting by providing a mark or letter for each size. The schedule also gives the reinforcing required. Of special interest to the carpenter is the pedestal size given in the schedule. The *pedestal* is a section of concrete smaller than the footing but placed at the same time, figure 12-15. This requires special forming. The purpose of the pedestal is to make the footing thicker where the column dowels are installed. This is needed sometimes to transfer the column load to the footing properly. The steel bars in the column might push through the footing if the footing is too thin.

SUMMARY

- Footings are pads of concrete which distribute building loads to the ground.

- Wall footings are usually continuous.

- Column footings are usually square.

- Footings need to be located concentric to the center line of the wall or column they support.

- Uneven pressure beneath a footing may cause it to fail by tipping.
- Retaining wall footings are not centered under the wall because of the lateral pressure on the wall caused by the supported earth.
- The thickness of wall footings depends upon the distance they project.
- Reinforced footings may project much farther than plain footings of the same thickness.
- Wall footing sizes may be given by dimensions or by notes.
- The elevation of the tops of the footings must be determined from the plans.
- A wall may have spaced footings or piers instead of continuous spread footings.
- Column footings are located by column center lines.
- Column footing sizes usually are given in a footing schedule.
- Column pedestals require special forming.

REVIEW QUESTIONS

1. What is the name of the part of a building foundation which rests on the ground and is constructed first?
   a. beam   c. footing
   b. slab   d. wall

2. The size of a footing depends upon all but one of the following. Identify it.
   a. load   c. soil type
   b. climate   d. materials

3. If column footings are not located under the center of the column,
   a. the footing may tip due to uneven pressure.
   b. the footing will spread the pressure out evenly.
   c. the footing will crack.
   d. no harm is done if the footing is the correct size.

4. A cantilever retaining wall has its footing offset because
   a. there is not enough space beside the wall to center the footing.
   b. it is easier to build the forms this way.
   c. the earth pushes against the side of the wall.

5. Sketch a continuous wall footing built on a sloping site.

6. Sketch a section of a typical unreinforced wall footing. Label the footing's dimensions assuming a 12″ thick wall.

7. Sketch a reinforced wall footing indicating where the bars are placed to strengthen them.

8. How are footings indicated on building plans?

9. How are the sizes of the following given on the plans?
    a. wall footings   b. column footings

10. Where is information about the elevation of the top of the footing found in the blueprints?

Unit 13
Foundation Walls

OBJECTIVES

After studying this unit, the student should be able to:

- identify two functions of foundation walls.

- name three reasons for forming special notches and pockets in foundation walls.

- tell how openings are provided for when constructing foundations.

IDENTIFICATION AND LAYOUT

Foundation walls rest on the footings and support the upper part of the building called the *superstructure.* They also may serve to enclose the basement. Large commercial buildings sometimes have two or three levels below ground. Most commercial buildings have poured concrete foundation walls. Concrete block foundations may be used for low walls.

The principle exterior layout lines for the building are the faces of the foundation walls. The carpenter begins the formwork by setting the outside form to these lines, figure 13-1. If the height and thickness of the wall can be read from the section view, little other information is needed from the plans to complete this formwork.

SPECIFICATION NOTES

The specifications contain information about the materials and *tolerances* (allowable deviations from true dimensions) required in the construction of the forms and should

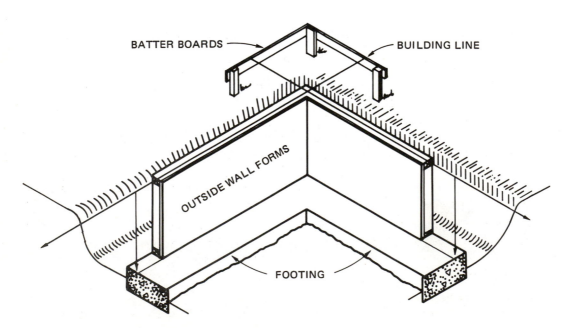

Fig. 13-1 Setting outside wall forms

be reviewed. Concrete formwork is found in Division Three of the C. S. I. Specification Format. It is one of several sections of this Concrete Division. The numbering system used varies but always contains the division number plus section numbers or letters. A five-digit numbering system in common use assigns Section 03100 to Concrete Formwork. This seems like a very large number, but it is really quite simple when understood. Figure 13-2 explains the meaning of these numbers and gives an example specification illustrating the C. S. I. 3-Part-Section Format.

SECTION 03100 CONCRETE FORMWORK

The first two digits, 03, indicate the Division where this Section is located, namely Concrete. Two digits are required because there are 16 Divisions.

The last three digits, 100, are used as a permanent identification for Concrete Formwork. Three digits are used in order to have plenty of room for subheadings, up to 99, under the broad heading of Concrete Formwork. For example, 03125 means Division 03 Concrete, Section 100 Formwork and subsection 25 Custom Steel Forms.

1. GENERAL

    1.01 DESCRIPTION
      A. Work included: form all cast-in-place concrete.
      B. Related work described elsewhere. Concrete reinforcement: Section 03200

    1.02 QUALITY ASSURANCE
      A. Provide at least one qualified person to be present at all times during this work.
      B. Comply with all pertinent recommendations contained in "Recommended Practice for Concrete Formwork" ACI 347.

2. PRODUCTS

    2.01 FORM MATERIALS
      A. Form lumber shall be Douglas Fir-Larch, number two grade, seasoned, surfaced four sides.
      B. Plywood, "Plyform" class I or II, DFPA.

3. EXECUTION

    3.01 CONSTRUCTION OF FORMS
      A. Construct forms substantially, mortar tight.
      B. Form all cast-in-place concrete to the sizes, shapes, and dimensions on the Drawings.
      C. Make proper provision for all openings, offsets, recesses, anchorage, and other features of the Work as shown or required.
      D. Construct all forms straight, true, plumb, and square within a tolerance horizontally of one in 200 and vertically of one in 500.

    3.02 PLYWOOD FORMS
      A. Nail plywood panels directly to studs in a manner to minimize the number of joints.
      B. Make all panel joints tight butt joints with all edges true and square.

    3.03 RE-USE OF FORMS
      A. Forms may be re-used four times unless damaged or rejected by the Architect.
      B. Re-use of forms shall in no way delay or change the best schedule for placement of concrete.

Fig. 13-2 C. S. I. specification notes for concrete formwork

Foundation walls would be quite simple to construct if it were not for the special details involved, such as tolerance, ledges, pockets and embedded items. The specifications spell out a list of specials in a few short sentences. Finding the necessary information on the drawings is more of a problem.

TOLERANCES

Tolerances and surface quality usually vary depending upon whether the concrete shows when the building is finished. For example, the form must be designed to limit the bulging between studs to 1 in 360 for exposed work, whereas a tolerance of 1 in 270 or 1 in 200 might be allowed for concealed concrete walls. The alignment of the walls is also controlled by tolerance specifications. These specifications are stated in 3.01 D, figure 13-2, page 119. Foundation walls may be partly exposed outside and do show inside if there is a basement. Other below-grade concrete foundations that are concealed can be rougher.

OFFSETS

Straight walls in foundations are probably the exception rather than the rule. Foundations must be shaped to fit the building above. Offsets such as projections, pilasters, recesses, angles, curves, and the like must be carefully duplicated in the forms. Figure 13-3 shows examples of foundation offsets.

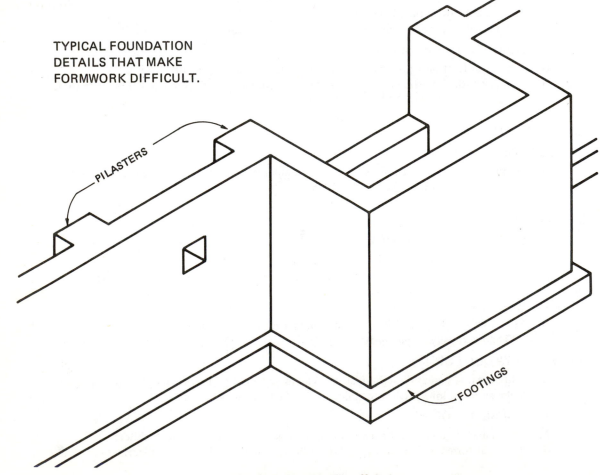

TYPICAL FOUNDATION
DETAILS THAT MAKE
FORMWORK DIFFICULT.

PILASTERS

FOOTINGS

Fig. 13-3 Foundation offsets

LEDGES AND POCKETS

Special notches formed in the wall frequently are needed to hold other materials in the proper place. When the notch is continuous and on the outside of the wall, it is called a *brick ledge*. It is formed by adding an insert to the regular wall form, figure 13-4. The brick ledge size shows on the wall section, but the foundation plan gives the location. The elevation of the bottom of the brick ledge is also shown on the plan. Usually the brick work extends to the ground line which may vary from place to place. The elevation sheets in the architectural subgroup should be checked to be sure the brick ledge is low enough at all points.

Sometimes a ledge is needed on the inside of the wall at the top. This is called a *joist ledge* and is formed to receive the ends of steel joists or small beams, figure 13-5. A similar ledge parallel to the joists also might be needed to support the edge of the slab, figure 13-6, page 122.

Joists are closely spaced members, 24 inches or less on center, so the continuous ledge is the easy way to hold them. When the structural members are farther apart, they are called *beams*. Notches called *beam pockets* are provided at the ends of these members, figure 13-7, page 122. These are not continuous, so their locations are critical. Center lines of columns help locate beams on the usual foundation plan.

OPENINGS

Window and door openings in basement walls must also be provided in the foundation.

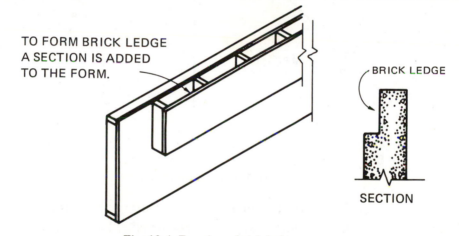

TO FORM BRICK LEDGE A SECTION IS ADDED TO THE FORM.

BRICK LEDGE

SECTION

Fig. 13-4 Forming a brick ledge

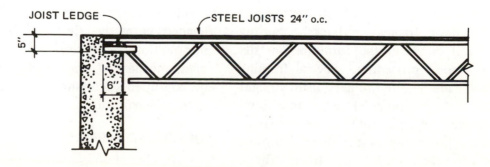

JOIST LEDGE

STEEL JOISTS 24" o.c.

5"

6"

Fig. 13-5 Joist ledge detail

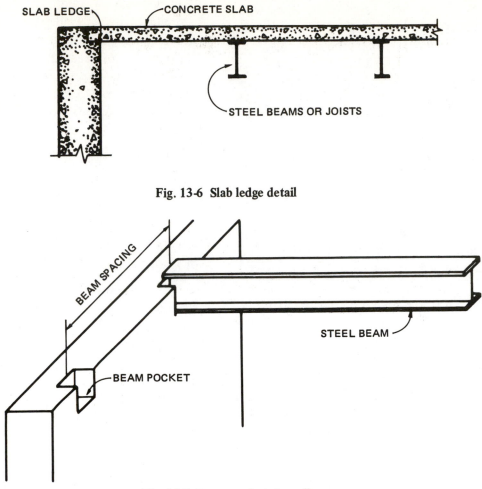

Fig. 13-6 Slab ledge detail

Fig. 13-7 Beam pockets in wall

The architectural details must be studied to understand what is required. Windows are usually installed after the wall is poured. Grooves or rough bucks are required for this later installation, figure 13-8.

Doors usually are made of metal hung in metal door frames or bucks. Setting these metal bucks in the forms before the pour is a good procedure. Special forming and bracing are required to keep them straight and free from buckling. The concrete flows into the back of the buck to anchor it to the wall. The swing of the door also must be correct when setting the frame, figure 13-9.

During the setting of foundation walls, openings or holes for use by other trades are formed by the carpenter. Special grooves in a wall, either vertical or horizontal, are sometimes required for piping or electrical wiring. These *chases* must be sized and located correctly, figure 13-10. Since this information is not shown on the foundation plan, mechanical and electrical plans must be consulted. Foremen in other trades also need to be consulted.

EMBEDDED ITEMS

Anchor bolts, inserts for pipe hangers and equipment, reglets, masonry anchors,

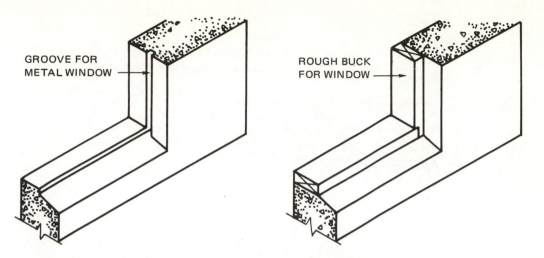

Fig. 13-8 Details for window openings

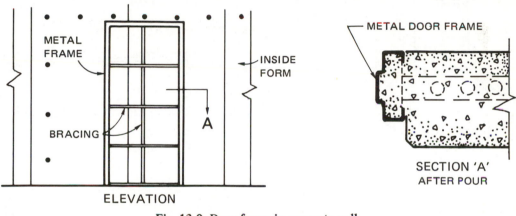

Fig. 13-9 Door frame in concrete wall

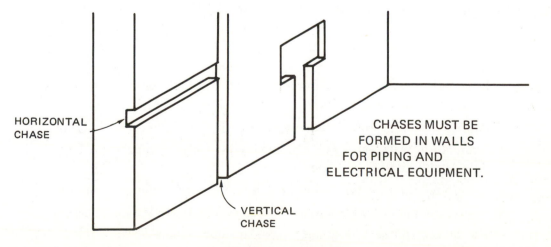

Fig. 13-10 Wall chases

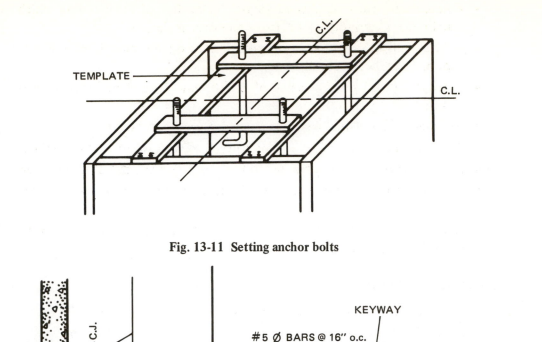

Fig. 13-11 Setting anchor bolts

Fig. 13-12 Construction joints

angles, weld plates, and other items also must be set. Anchor bolts are set with precision by using templates of the bolt layout, figure 13-11. Construction joints between pours require keyways and dowels, figure, 13-12.

SUMMARY

- Foundation walls support the upper part of a building by transferring the load to the footings.
- Poured concrete foundations are typical for commercial buildings.
- Foundation walls are formed by setting the outside form to the layout lines established on the batter boards.
- The foundation section gives sufficient data for forming most walls.
- Information about materials and tolerances is found in the specifications.
- Tolerances and surface quality vary depending upon whether the concrete will be seen in the finished building.

- Foundations must be shaped to fit the outline of the building which usually means a number of offsets, recesses, angles, etc.
- A brick ledge is a special continuous notch on the outside of a wall to support brick work usually at the ground line.
- A ledge formed on the inside of a wall may be used to support joists or small beams.
- Notches for individual beams are called beam pockets.
- Window and door openings are formed according to section details.
- Steel door frames are often set in the forms and anchored by the pour.
- Openings for mechanical and electrical work must be formed by the carpenter.
- Embedded items require special care in setting because they cannot be moved after the concrete is poured.

REVIEW QUESTIONS

1. The foundation walls rest upon the footings and support the
   a. basement floor.
   b. the interior columns.
   c. the superstructure.

2. Most commercial buildings have foundation walls constructed of
   a. concrete blocks.   c. stone.
   b. bricks.   d. poured concrete.

3. Concrete formwork specifications are found in which Division of the C. S. I. Format?
   a. Division 3   c. Division 8
   b. Division 6   d. Division 16

4. Concrete in buildings with much exposed concrete work is classed as
   a. mass concrete.   c. structural concrete.
   b. architectural concrete.   d. lightweight concrete.

5. When do foundation walls show inside the building?
   a. never   c. when there is a basement
   b. when built on a sloping site d. when built of brick

6. Continuous notches on the outside of a foundation wall are called
   a. brick ledges.   c. slab ledges.
   b. joist ledges.   d. beam pockets.

7. Joists are closely spaced members with a maximum spacing of
   a. 16 inches.   c. 24 inches.
   b. 18 inches.   d. 36 inches.

8. Special grooves in walls for piping or wiring are called
   a. dados.   c. reveals.
   b. returns.   d. chases.

9. How are window openings in foundation walls prepared to receive metal windows?

10. Metal door frames are best set in the forms and anchored by the concrete pour. Sketch a section of a metal door frame set in a concrete wall.

11. Sketch a construction joint as used in foundation walls.

12. Sketch a section of foundation wall with a brick ledge and a footing.

Unit 14
Special Foundations

OBJECTIVES

After studying this unit, the student should be able to:

- describe various special foundations and tell how they are shown on the plans.
- state uses of the various special foundations.
- tell how frost in the ground affects foundations of buildings.

PIER FOUNDATIONS

When a building does not have a basement, the foundation system depends upon several factors. If the soil is not of suitable density for several feet below the building, pier foundations are used. The most common type is the round drilled pier. This type pier is shown on the foundation plan as a circle beneath the wall. A situation may arise that requires a pier of greater bearing capacity. A flared or bell bottom pier is used. This type appears on the plan as two circles, one a bit larger than the other, figure 14-1. Square piers can be used, but they are more expensive when deep in the ground because there is no equipment to drill a square hole.

Piers are spaced several feet apart so there must be something above them to carry the walls and floors. Reinforced concrete beams generally are used when in contact with the ground. These beams are called *grade beams* because they occur at the grade or ground line, figure 14-2, page 128. They appear on the drawing much like a foundation wall but have some identification such as GB-1, GB-2, etc., figure 14-3, page 128.

Piers also can be used under interior walls and slabs with supporting grade beams. Floor slabs sometimes are supported by a pattern of piers, figure 14-4, page 128. This is done when there is considerable fill under the slab and thorough compaction is too

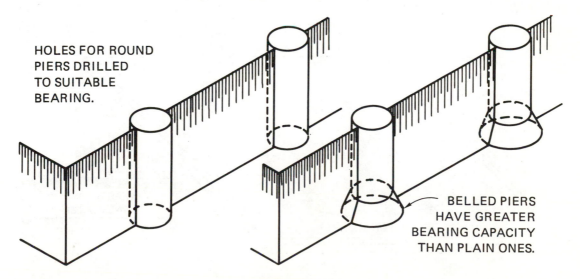

HOLES FOR ROUND PIERS DRILLED TO SUITABLE BEARING.

BELLED PIERS HAVE GREATER BEARING CAPACITY THAN PLAIN ONES.

Fig. 14-1 Round drilled piers

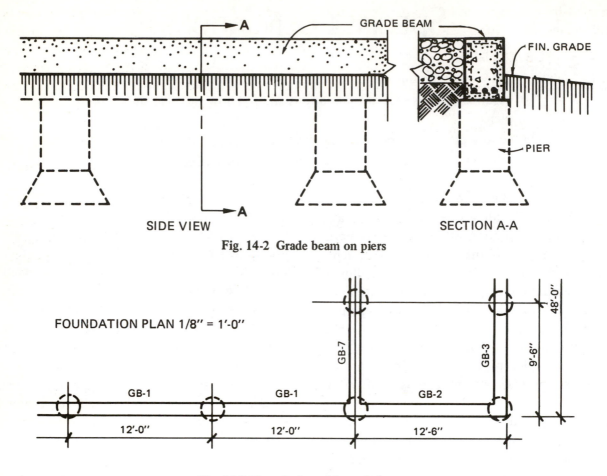

SIDE VIEW   SECTION A-A

Fig. 14-2 Grade beam on piers

Fig. 14-3 Foundation with grade beams

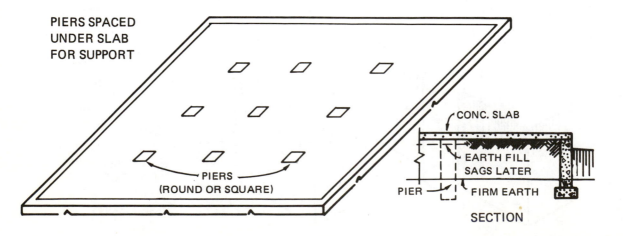

Fig. 14-4 Floor slab on piers

expensive. After a time the fresh fill under the slab sinks a bit, but the slab will be supported by the piers. In this system, the earth fill serves as a form for the concrete slab rather than as a permanent support.

PILE FOUNDATIONS

A *pile* is a long slender support made of timber, concrete, or steel driven into the ground. Pile foundations support their loads by means of many piles driven deeply into the ground. Some are driven to solid rock and thus support the load like posts or columns. When the rock layer is too deep, the piles support the load by friction along their sides, figure 14-5. The foundation plan shows these piles, usually in groups or clusters, beneath columns and grade beams, figure 14-6. The carpenter forms the pile cap which covers each group of piles. Usually there are anchor bolts and reinforcing steel to be set. The elevation of the top of the pile cap, (as well as the exact location of anchor bolts) must be established carefully.

STEEL-GRILLAGE FOUNDATIONS

For very heavy column loads in multi-story steel framed buildings, the steel-grillage foundation is used. Here the column load is spread over a considerable area by short steel beams, figure 14-7, page 130. Normally steel columns rest on a base plate called a steel

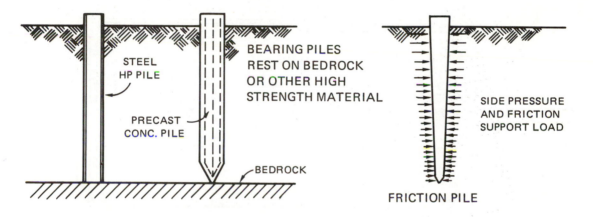

Fig. 14-5 Foundation piles

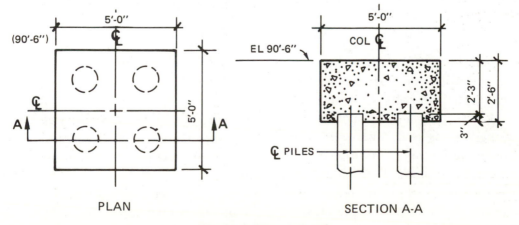

Fig. 14-6 Pile cap details

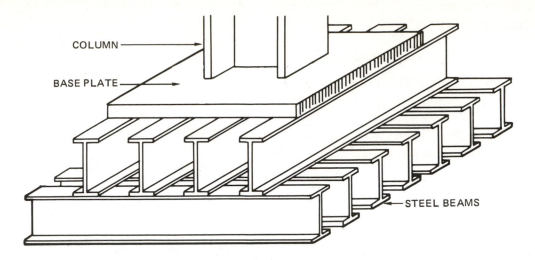

COLUMN

BASE PLATE

STEEL BEAMS

Fig. 14-7 Steel grillage foundation

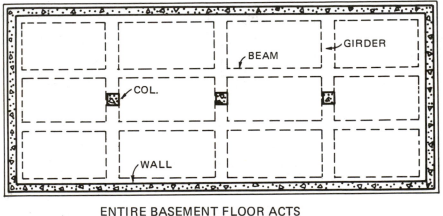

GIRDER

BEAM

COL.

WALL

ENTIRE BASEMENT FLOOR ACTS
AS FOOTING. BEAM AND GIRDERS
CAST WITH FLOOR DISTRIBUTE LOAD.

Fig. 14-8 Mat foundation

billet which spreads the load to the concrete. The billet also provides anchorage for the column through the anchor bolts. However, when the load is great, the plates are not enough, and the steel grillage is used. The steel grillage rests on a concrete footing or pile cap. A grout space between the grillage and the concrete is required to level up the assembly. The elevation of the top of the concrete must be read from the plans and accurately measured.

MAT FOUNDATIONS

Buildings built on very deep deposits of low-density soil, or on uneven subsoil, may be supported by a mat or floating foundation. An example of a mat foundation is an entire basement floor which is designed as a spread footing. It may be a single thick slab or a system of beams and slabs. In this case the foundation plan looks like a reinforced concrete floor plan, figure 14-8. No footings, piers, or piles are used, which makes this foundation plan unusual.

COMBINED FOUNDATION FOOTINGS

A combined foundation footing design is used when the building is placed flush along a property line. In this case, footings for exterior columns cannot be centered under the columns. If they were, they would extend into the next property, figure 14-9. This problem can be solved with the combined footing which joins the exterior footing with the next interior footing, figure 14-10. By careful proportioning of the footings,

the loads can be spread uniformly to the soil. The tendency to tip is resisted by the reinforced concrete connecting element.

WOOD FOUNDATIONS

A new idea in foundations for light construction is the all-weather wood foundation, figure 14-11, page 132. No concrete footing or wall is used. Wall panels are fabricated of pressure-treated lumber and plywood and quickly erected on 2 x 8 treated footing plates. The footing plates rest on a carefully leveled gravel fill below the frost line. Its use drastically reduces construction

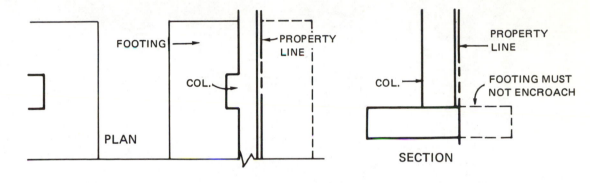

Fig. 14-9 Footing at property line

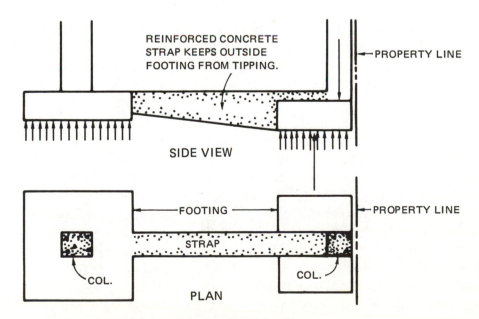

Fig. 14-10 Cantilever combined footing

Fig. 14-11 All-weather wood foundation

delays due to bad weather including winter cold. If it can be used, the wood foundation makes a warmer, drier, more livable basement when insulated properly, figure 14-12.

FROST DEPTHS

When soil freezes, it expands because of the ice crystals that form. Frozen ground is hard and strong but unstable when it thaws. Arctic structures can be built on *permafrost* (frozen ground that never thaws). The heat of the building, however, must not be allowed to reach the ground, so thick insulation is required beneath the foundation and floors. In the United States average frost depths vary from 0 to 72 inches, figure 14-13, page 134.

Most foundations are carried below the local frost depth. Local building codes usually specify the minimum depth of foundation footings. Specifications also include a statement that requires footing trenches to be dug to undisturbed firm soil of adequate bearing capacity.

Grade beams bear on piers that extend to solid bearing. Between the piers, the grade beam is made deep enough to carry the loads and reach below the frost line. In some soils, those which expand and contract due to moisture changes, special details, such as compressible fiber inserts, are required under the grade beam, figure 14-14, page 135. This is done to keep the periodic heaving of the soil from lifting the grade beam.

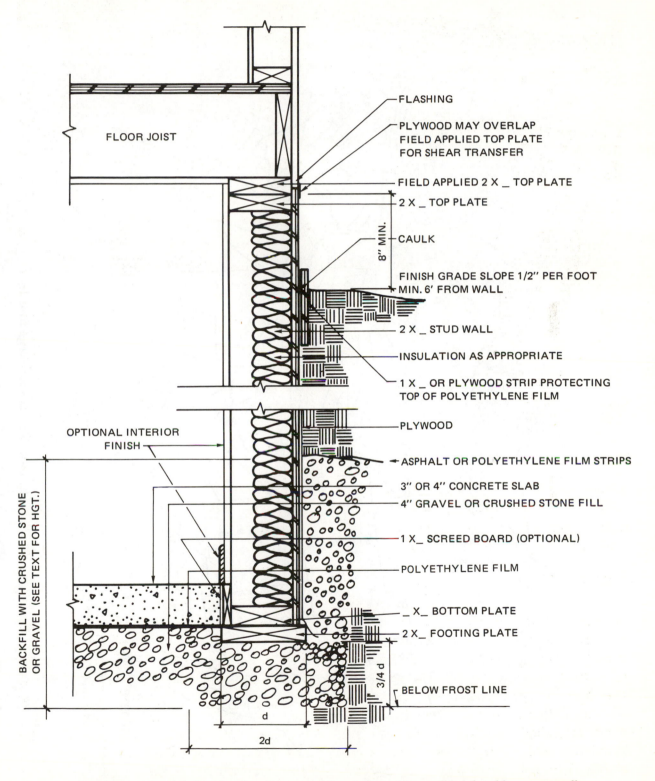

FLASHING

PLYWOOD MAY OVERLAP
FIELD APPLIED TOP PLATE
FOR SHEAR TRANSFER

FIELD APPLIED 2 X _ TOP PLATE

2 X _ TOP PLATE

CAULK

FINISH GRADE SLOPE 1/2'' PER FOOT
MIN. 6' FROM WALL

2 X _ STUD WALL

INSULATION AS APPROPRIATE

1 X _ OR PLYWOOD STRIP PROTECTING
TOP OF POLYETHYLENE FILM

PLYWOOD

ASPHALT OR POLYETHYLENE FILM STRIPS

3'' OR 4'' CONCRETE SLAB

4'' GRAVEL OR CRUSHED STONE FILL

1 X_ SCREED BOARD (OPTIONAL)

POLYETHYLENE FILM

_ X_ BOTTOM PLATE

2 X _ FOOTING PLATE

BELOW FROST LINE

8'' MIN.

3/4 d

FLOOR JOIST

OPTIONAL INTERIOR
FINISH

BACKFILL WITH CRUSHED STONE
OR GRAVEL (SEE TEXT FOR HGT.)

d

2d

Fig. 14-12 A section drawing showing a wood foundation with a well-insulated basement wall

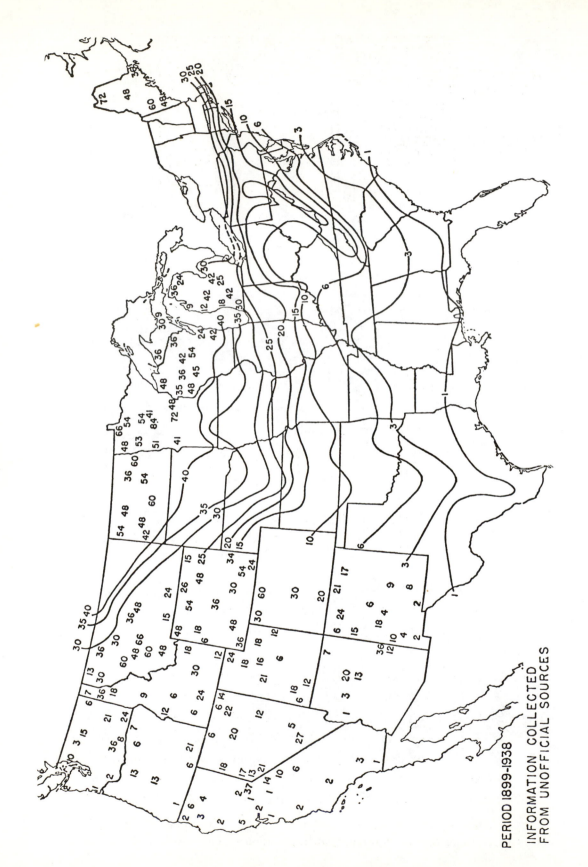

PERIOD 1899-1938

INFORMATION COLLECTED
FROM UNOFFICIAL SOURCES

Fig. 14-13 Map showing average frost depths

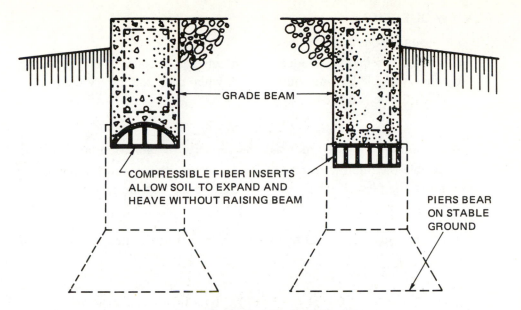

Fig. 14-14 Grade beam inserts

SUMMARY

- Pier foundations are used when the soil is of poor density and there is no basement.

- Flared piers are able to carry greater loads than straight ones.

- Grade beams are used to transfer the building loads to the piers.

- Pier supported floor slabs are poured on fill which acts as a form rather than as a permanent support.

- Piles are long slender supports driven into the ground.

- Piles may support loads by bearing on rock or by friction along their sides.

- Concrete pile caps are formed by the carpenter.

- Steel grillage foundations are used for very heavy column loads.

- A grout space is required under column base plates and steel grillage beams.

- Mat foundations are pads as large as the building and are used in soils of low density.

- Combined footings are used when exterior footings cannot be centered under the columns because of property rights.

- Wood foundations made of pressure treated lumber and plywood can be used in place of concrete in light construction if properly constructed.

- Frozen ground is unsuitable for foundations if it will thaw later.

- Building codes specify the required depth of foundations in order to place them below the local frost line.

REVIEW QUESTIONS

1. Pier foundations are used
   a. when a basement is needed.
   b. when suitable bearing soil is close to the surface.
   c. when poor soil is several feet deep.
   d. when poor soil is very deep.

2. An economical way to increase the capacity of pier foundations is to:
   a. double the number of piers.
   b. make the piers deeper.
   c. increase the size of the piers.
   d. flare the bottom of the piers.

3. Why are square piers more expensive than round ones when deep in the ground?

4. The reinforced concrete members that carry loads to the piers are called
   a. joists.
   b. grade beams.
   c. girders.
   d. stringers.

5. Sketch part of a foundation showing grade beams and round flared piers.

6. Name two ways foundation piles support loads.

7. Piles are driven in groups of two or more and each group is covered with
   a. a concrete cap.
   b. a steel billet.
   c. a steel grillage.
   d. a wooden mat.

8. Steel grillage foundations are used when
   a. piles are required.
   b. very heavy column loads must be carried.
   c. pier foundations are not practical.
   d. a shallow foundation is desired.

9. What is the name of the foundation system that has no footings, no piers, and no piles but supports sizable buildings on low density soil?

10. Why is it necessary to use a combined footing for a column built against a property line?

11. What kind of wood is used to construct the all-weather wood foundation?

12. Name two factors that affect the minimum depth of footings for the typical foundation.

13. Draw and label a sketch showing how grade beams are constructed to resist periodic heaving of very expansive soils.

Section IV
Wood Framing Work

Unit 15
Floor Systems

OBJECTIVES

After studying this unit, the student should be able to:

- describe three different wood floor framing systems.
- sketch two joist-to-girder framing sections.
- give two ways used on drawings to show joist framing information.

JOIST AND GIRDER SYSTEM

For crawl space buildings, wood floor structural systems are shown on the foundation plan. The type of note typically used is given in figure 15-1. The direction of span, size of joist, and joist spacing are given. The location and size of girders are indicated also. The foundation sections show the structural framing and how it relates to the masonry materials, figure 15-2. Note the different heights required for the piers and foundation to receive the wood members properly.

Another method of drawing this structural data is the framing plan. Its use is limited in wood joist construction but common for heavier structures. The floor structure is shown from above before the subflooring is installed. Joists, headers, blocking, bridging, and openings are clearly shown, figure 15-3.

WOOD GIRDER

The built-up girder shown in figure 15-2 has two advantages. The most important advantage is the greater headroom or clearance provided under the girder when installed as shown. In crawl space buildings there is minimum space (18 in. sometimes) below the joists. If the girder is not built-in flush with the tops of the joists, it projects down into this space. Piping and air-conditioning ducts are easier to install if the girder is held up. The second advantage for this arrangement is the equalization of shrinkage of material over the whole floor system. When joists run over the wood girders, the depth of wood is approximately doubled, figure 15-4. Even with kiln dried lumber, seasonal changes in moisture content can cause troublesome movement in the structure

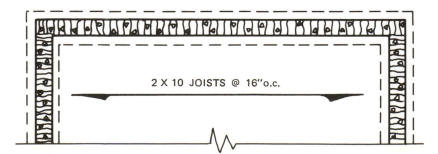

2 X 10 JOISTS @ 16"o.c.

Fig. 15-1 Joist note on foundation plan

138

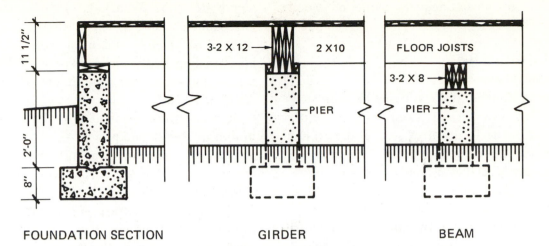

FOUNDATION SECTION   GIRDER   BEAM

Fig. 15-2 Foundation sections

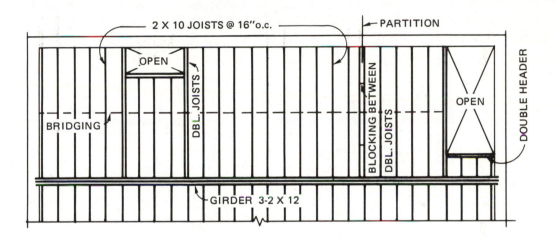

Fig. 15-3 Floor framing plan

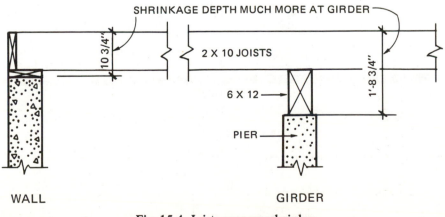

WALL   GIRDER

Fig. 15-4 Joists over wood girder

because of this unequal depth of wood material.

STEEL GIRDER

Steel beams are sometimes used as girders in wood framed floor systems because steel does not shrink and swell with moisture changes. The location of steel girders is not as critical as for wood girders. Figure 15-5 shows the steel beam below the wood joists. A nailing plate is bolted to the top of the steel member. With the joists on top of this girder, the carpenter has two options for making the splices. The joists may lap each other which is quite simple to do. The joists do not have to be cut to an exact length, and when nailed together the joist spans are effectively tied together. The second option is to butt the joists over the girder and splice with a short length of wood or metal. In this case the joists must be cut to exact lengths. An advantage of this method of splicing with the joists in line is that the plywood subflooring can be installed with a minimum of cutting. Another advantage of joist butting into beams is that more headroom is allowed for installation of ductwork and piping.

When headroom is a problem, the steel girder can be detailed partly within the joist depth, figure 15-6. The top of the wood joists should be at least 1/2 inch above the top of the steel girder. This is to prevent the steel beam from making a hump in the floor should the wood joists shrink after the building is heated. Metal straps or the plywood subflooring tie the joists together. Nailing blocks should be bolted to the steel member as well.

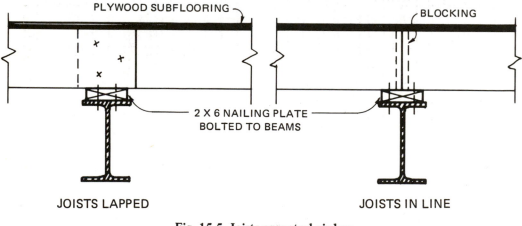

Fig. 15-5 Joists over steel girders

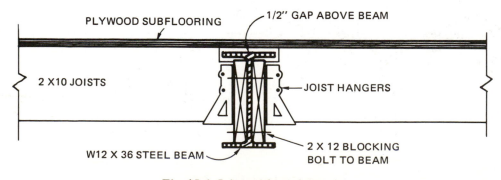

Fig. 15-6 Joists with steel girder

SPECIFICATIONS AND DETAILS

Specifications must be checked for lumber grades, bridging requirements, framing of openings, and double joists. Double joists are required under partitions running parallel to the joists. They usually are spaced and accurately located so that pipes and electrical wiring are easy to install, figure 15-7.

Location of openings for stairs, fireplaces, chimneys, chutes, etc. must be done from information on several sheets. Details of stairs and fireplaces give useful information. The overall floor plan should be checked for additional information.

2-4-1 PLYWOOD SYSTEM

A simplified wood floor system developed by the plywood industry consists of 1 1/8" plywood over 4 x 6 girders. No joists are used when the girders are spaced 4'-0" o. c. (on center). This system is indicated on the foundation plan and explained in the details, figure 15-8. One of the benefits of this system is the lowering of the floor line by about 6 inches. Labor costs are also less than with the conventional joist system.

TRUS JOIST SYSTEM

For long-span floor and roof framing, a patented assembly made of wood *chords*

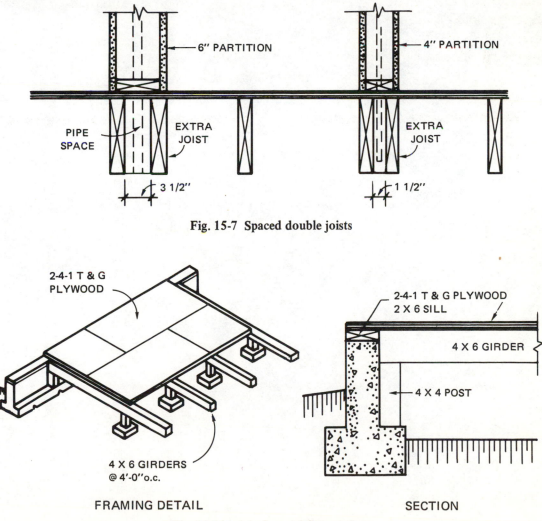

Fig. 15-7 Spaced double joists

FRAMING DETAIL

SECTION

Fig. 15-8 2-4-1 plywood floor system

(top and bottom members) and metal diagonals is available. The wood chords provide nailing surfaces for decking and ceiling finish, figure 15-9. Because of their special fabrication, shop drawings are usually required. A framing plan shows the location of each joist. Anchorage of the joist to the structure is shown in the details, figure 15-10.

PLANK FLOOR SYSTEMS

Used often with heavy timber and glue-lam structural framing (Unit 18), plank floor and roof decking is another system installed by the carpenter. Planking of nominal 2, 3, and 4 inch thicknesses is available, figure 15-11, page 144. The advantages of this system are the wider spacing of beams and the reduced depth of construction.

For heavy floor loads the laminated floor is a proven system, figure 15-12, page 144. Planks are turned on edge like joists but are nailed one against the other to form a solid slab of wood. A wearing surface of hardwood flooring is usually applied over the laminated deck.

Fig. 15-9 Trus joist framing system

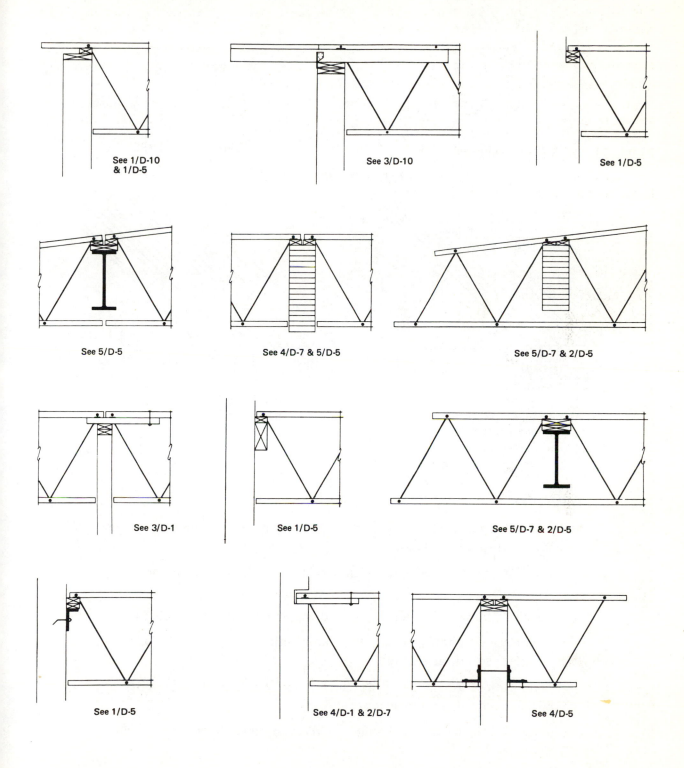

Fig. 15-10 Trus joist bearing details

4 X 6

3 X 6

2 X 6

SOLID LUMBER TYPES   LAMINATED TYPES

Fig. 15-11 Heavy timber decking

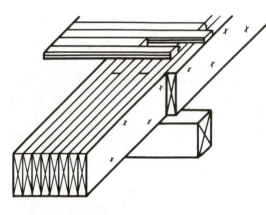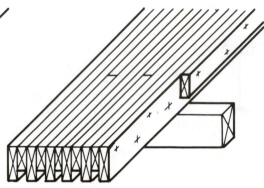

HEAVY MILL CONSTRUCTION,
2 X 8's WITH HARDWOOD
WEARING SURFACE.

ALTERNATING 2 X 6 AND
2 X 4 CONSTRUCTION. MAY
BE USED EXPOSED FOR
FLUTED APPEARANCE.

Fig. 15-12 Laminated wood decking

SUMMARY

- Sizes of wood joists and girders are given by notes on foundation plans for crawl space buildings.

- Framing plans showing all members are sometimes drawn for wood structures.

- Specifications give lumber grades, bridging, and double joist requirements.

- Wood girders should be installed flush with the top of floor joists to minimize shrinkage effects.

- Steel girders should be set 1/2 inch below the top of floor joists to allow for shrinkage of new joists.

- Joists should be spliced in-line at girders to make plywood joints line up right.

- The 2-4-1 plywood system uses 1 1/8'' plywood over 4 x 6 girders at 4'-0'' o. c.

- TRUS JOIST systems are patented long span assemblies with wooden top and bottom chords and metal diagonals.

- Plank and laminated floor systems are used in heavy timber type construction for long spans and minimum floor thicknesses.

REVIEW QUESTIONS

1. Which of the following is not given in the joist framing note found on foundation plans?

   a. length of joists.   c. spacing of joists.
   b. size of joists.   d. direction of span.

2. A drawing showing all the joists in their correct arrangement is called a

   a. floor plan.   c. framing plan.
   b. section view.   d. pictorial drawing.

3. Give two advantages gained by framing a wood girder flush with the top of the joists.

4. When wood joists frame into the side of a steel girder, the top of the steel member should be

   a. flush with the top of the joists.
   b. about 1/2 inch below the top of the joists.
   c. at least 2 inches below the top of the joists.
   d. flush with the top of the subflooring.

5. Where is the information found that tells how the joists frame into the girder?

   a. sections.   c. specifications.
   b. foundation plan.   d. elevations.

6. Sketch two joist-to-girder framing sections.

7. Why are double joists under partitions often spaced a few inches apart?

8. Name two wood floor systems that do not use conventional joist framing.

9. Plank floor systems use thick lumber for the subflooring. What are three thicknesses available for these floors?

10. For heavy loads and long spans, a laminated wood floor can be used. Sketch a section of laminated floor construction of 2 x 8 lumber.

Unit 16
Walls

OBJECTIVES

After studying this unit, the student should be able to:

- explain the basic features of platform and balloon framing systems.
- tell how openings are dimensioned on the drawings for wood framed structures.
- describe three methods used to brace frame walls against wracking.

PLATFORM FRAMING

Platform framing, sometimes called western framing, is the basis for most wall framing today. It is an efficient, convenient, and safe method. Both exterior walls and interior walls (partitions) are framed the same, figure 16-1. Frame walls are drawn on the plans by means of symbols rather than

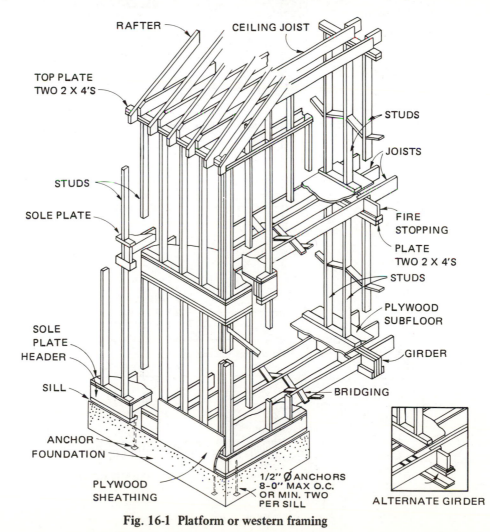

RAFTER

CEILING JOIST

TOP PLATE
TWO 2 X 4'S

STUDS

STUDS

JOISTS

SOLE PLATE

FIRE
STOPPING

PLATE
TWO 2 X 4'S

STUDS

PLYWOOD
SUBFLOOR

SOLE
PLATE
HEADER

GIRDER

SILL

BRIDGING

ANCHOR
FOUNDATION

PLYWOOD
SHEATHING

1/2" Ø ANCHORS
8-0'' MAX O.C.
OR MIN. TWO
PER SILL

ALTERNATE GIRDER

Fig. 16-1 Platform or western framing

by showing each stud (Unit 2 and Appendix). The architect assumes that the carpenter knows standard framing methods. The wall section, figure 16-2, is used to explain the completed wall. The size and spacing of studs and the plates required are included. Of

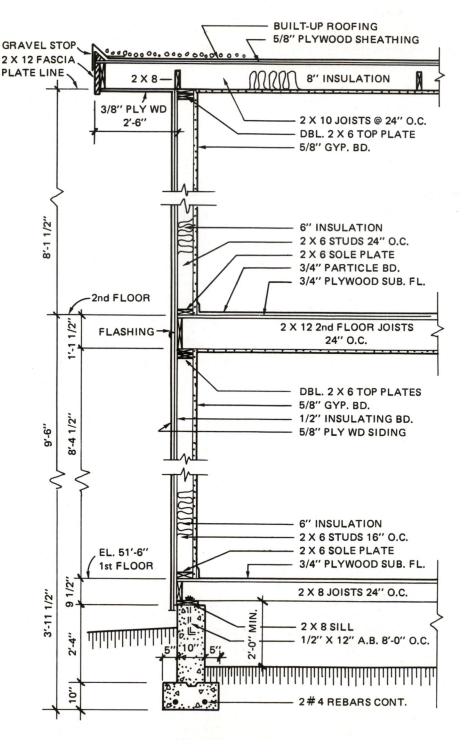

Fig. 16-2 Wall section 3/8″ = 1′-0″

special importance is the height of the structural assembly which is given on the wall section and on the elevations. The points dimensioned are the floor line and the next higher plate line. It is helpful if the floor line shown is the top of the subfloor in wood framed floor units. For concrete floors there is no problem since the bottom plate rests directly on the slab, figure 16-3. The subfloor also makes a working deck on which to put up the next set of walls. This is also a good safety factor.

BALLOON FRAMING

Balloon framing, also known as eastern framing, was invented before platform framing, figure 16-4, page 150. It is no longer as popular as platform framing but is often used in conjunction with the platform method.

A feature of balloon framing is its exclusive use of 2 x 4 studs. The studs run from the sill to the highest plate. For two-story buildings this means many 18 to 20 feet long studs. Even though studs this size are expensive and not readily available, they do minimize wood shrinkage in the total height of the building. Second floor joists rest on a

ribbon board notched into the studs, figure 16-5, page 151. The balloon frame ribbon board is still used when framing modern split-level floors, figure 16-6, page 151. Special architectural sections usually are drawn to give the carpenter the necessary information for these cases.

MOD 24 FRAMING

The most recent framing method developed is called Mod 24 framing. In an effort to conserve material and labor, 24-inch spacing of studs is used. Floor and roof framing at 24-inch spacing also contribute to the savings. Less labor is required because there are fewer pieces to handle and install.

Structurally, the 24-inch spacing of studs is adequate and approved by all model building codes for single-story construction. Research indicates that 24-inch framing can be used in two-story building and is gaining acceptance in single family and multifamily dwellings. The Mod 24 system works best when floor joists, studs, and roof trusses are aligned over each other. Structurally, the system produces a series of in-line frames using both lumber and plywood to best advantage, figure 16-7, page 151.

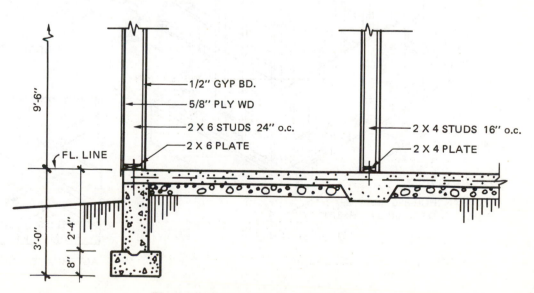

Fig. 16-3 Frame walls on concrete floor

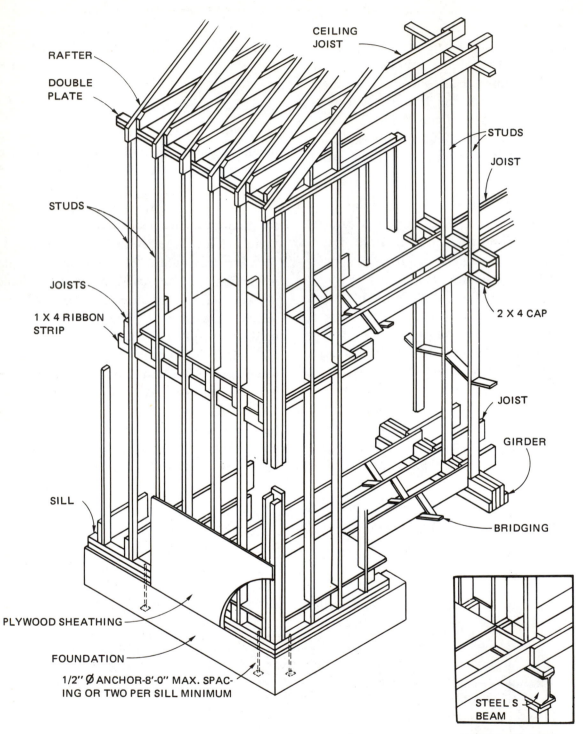

RAFTER

DOUBLE PLATE

CEILING JOIST

STUDS

JOIST

STUDS

JOISTS

1 X 4 RIBBON STRIP

2 X 4 CAP

JOIST

GIRDER

SILL

BRIDGING

PLYWOOD SHEATHING

FOUNDATION

1/2″ Ø ANCHOR-8′-0″ MAX. SPACING OR TWO PER SILL MINIMUM

STEEL S BEAM

ALTERNATE GIRDER

Fig. 16-4 Balloon framing

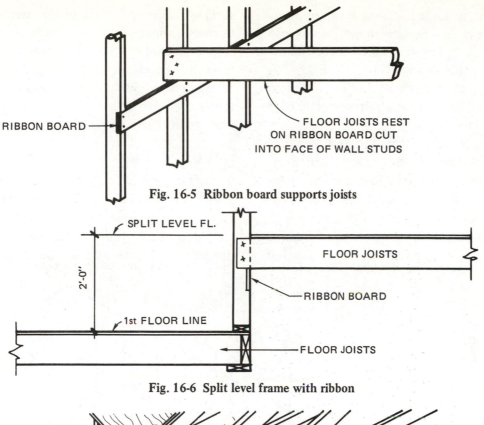

RIBBON BOARD

FLOOR JOISTS REST
ON RIBBON BOARD CUT
INTO FACE OF WALL STUDS

Fig. 16-5 Ribbon board supports joists

SPLIT LEVEL FL.

FLOOR JOISTS

2'-0"

RIBBON BOARD

1st FLOOR LINE

FLOOR JOISTS

Fig. 16-6 Split level frame with ribbon

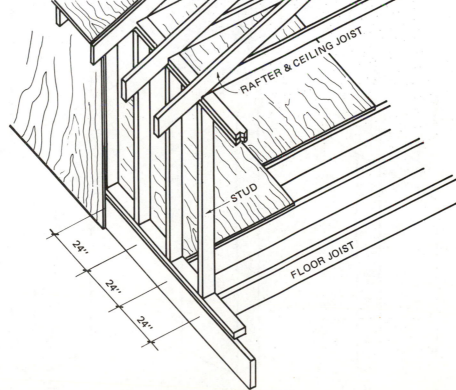

RAFTER & CEILING JOIST

STUD

FLOOR JOIST

24"

24"

24"

Fig. 16-7 Mod 24 in-line framing

Even more lumber can be saved by using single wall plates. Splices in wall and partition plates are made with metal plates manufactured for this purpose, figure 16-8. The single plate system requires careful placement of trusses and second floor joists directly over the studs below.

A further development of the Mod 24 system is known as the Arkansas Energy-Saving system. This system is primarily designed to cut energy costs. It uses 2 x 6 studs at 24-inch spacings in order to have 6-inch insulation in the exterior walls. Wall framing is simplified to reduce the amount of wood surface exposed to outdoor temperatures. Corners framed with two studs permit overlapped insulation and use clips to hold the drywall. Buildings built using all the energy saving techniques of this system have energy requirements reduced by 40% or more. Details are in figure 16-10.

WALL OPENINGS

Frame walls and openings are dimensioned on the floor plan from the exterior face of the outside walls. Usually windows and doors are dimensioned in a string of dimensions just outside the building (Unit 6).

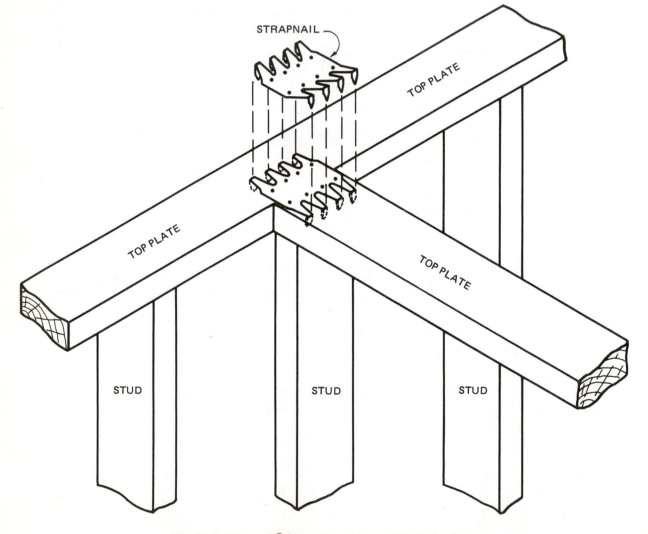

Fig. 16-8 Strapnail® fasteners used to splice single plates

The rough opening is laid out from its center line. This opening is calculated or given in the manufacturer's literature.

The next detail of openings is that of the header. The headers over openings must be designed to carry the loads from above, figure 16-9. They may be scheduled on the drawings or described by the specs. If neither source gives this information, the carpenter must check the building code or some other reliable source. Extra large openings require structural analysis. The Arkansas Energy-Saving framing uses one 2 x 6 laid flat over the opening with truss work or plywood box beams rather than multiple headers. Reasons for this are to save lumber and provide space for more insulation above the opening, figure 16-10.

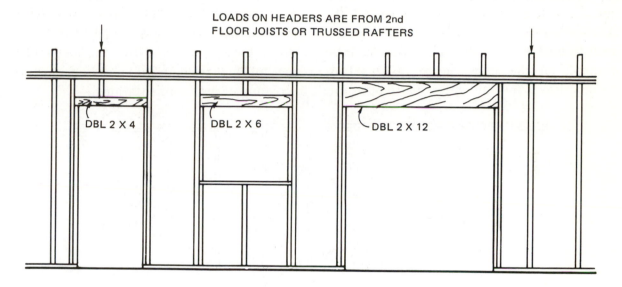

Fig. 16-9 Headers sized for openings

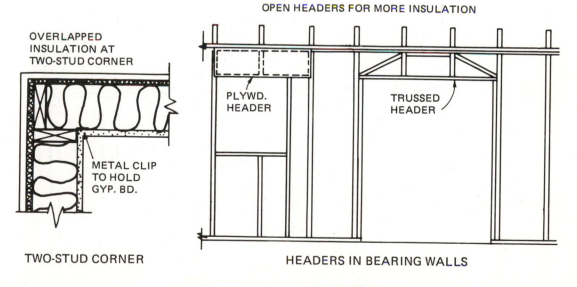

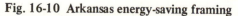

Fig. 16-10 Arkansas energy-saving framing

BRACING

Several methods are used to brace the walls against *wracking* (falling apart) due to windloads and earthquake loads. The specs or detail drawings usually state the method required. One of the oldest methods consists of a line of braces cut in between the studs. This forms a brace from sole plates to the top of the corner posts. A corner brace should be placed at each end of every exterior wall, figure 16-11. Another common method is the let-in brace. It is formed by notching a 1 x 6 brace into the face of the studs, figure 16-12. This method leaves the stud spaces more open for wiring and insulation than does the cut-in bracing of figure 16-11.

Other methods of bracing are those using wood sheathing or plywood sheathing. Wood sheathing applied diagonally to the wall gives good bracing but is wasteful of material. Plywood sheathing, however, when properly nailed, is an excellent way to brace the walls. Recent earthquakes have proved its value.

A very popular method of corner bracing is to use one sheet of plywood at each end of the wall and insulating sheathing over the rest of the surface, figure 16-13. This method has the bracing strength of plywood and the economy of insulating sheathing.

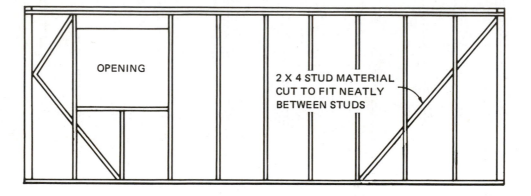

Fig. 16-11 Cut-in corner braces

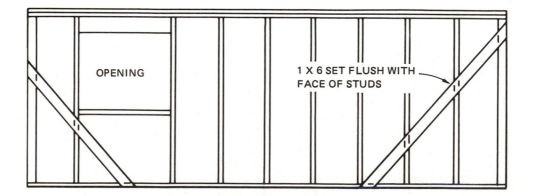

Fig. 16-12 Let-in corner braces

Fig. 16-13 Plywood corner bracing

SUMMARY

- Western platform framing is now widely used in the United States.

- Frame walls are drawn by means of symbols on the plans.

- Wall sections give much of the data needed to frame walls.

- Balloon framing details are still used in some split-level frames.

- Mod 24 framing is a new framing method in which studs are spaced 24 inches o.c.

- Window and door openings are dimensioned to their center lines on the floor plans.

- Headers over openings must be sized to carry the loads resting on them.

- Bracing is required to brace walls against forces caused by wind and earthquakes.

- Plywood sheathing at the corners with insulating sheathing between is a popular method for bracing walls.

REVIEW QUESTIONS

1. The most important framing method in use today is called
   a. eastern braced framing.
   b. balloon framing.
   c. western platform framing.
   d. post and beam framing.

2. Frame walls are shown on the floor plans by
   a. a single heavy line for each wall.
   b. double lines spaced according to the thickness of the wall.
   c. drawing every stud in its proper place.

3. Which type drawing gives the size, spacing, and height information for framing a wall?

4. Which points are dimensional on a wall section?

5. Sketch the ribbon board detail used to support second floor joists in the balloon frame.

6. Why is the platform framing system safer to erect than the balloon frame?

7. How does the 24-inch spacing of members in Mod 24 framing reduce the cost of construction?

8. Why are 2 x 6 wall studs used when building houses by the Arkansas Energy-Saving standards?

9. How are openings in frame walls dimensioned on a floor plan?

10. Which of the following corner bracing methods is very popular now?
    a. let-in brace
    b. cut-in brace
    c. diagonal sheathing
    d. plywood at corners

11. Study figure 16-2, Wall Section, and answer the following questions:

    a. What is the scale of this drawing?

    b. Is the height of the wall drawn full-scale or broken?

    c. What is the elevation of the first floor?

    d. Is the elevation of the first floor at the finish floor line or at the subfloor line?

    e. What is the elevation of the second floor?

    f. If 2 x 6 lumber is actually 1 1/2 inches thick, what are the stud lengths for first and second story walls?

    g. What is the spacing of first floor studs?

    h. What material is used for inside finish?

    i. How thick is the insulation in the walls and in the roof?

    j. What type framing is indicated in figure 16-2?

Unit 17
Roofs

OBJECTIVES

After studying this unit, the student should be able to:

- identify by naming and drawing common types of roofs.
- explain how the slope of roofs is specified on drawings.
- describe trussed rafters and tell how they are used.

ROOFS CLASSIFIED

Roofs framed of wood are probably the most interesting part of carpentry construction. There are many different types of roofs. All are similar in that they are made of many small members arranged in repeating patterns. The design of a roof framing system is best viewed before the sheathing and shingles are placed, figure 17-1.

Two general classes of roofs are flat or low-sloped roofs, and medium to steep-sloped roofs. The main reason for dividing roofs into these two classes is the kind of roofing used on each. The flat roof requires a built-up roofing system while the steep roof usually

has shingles. A 1/8 pitch roof (3" rise per 12" run) is about the minimum pitch for a roof with shingles. Specifications sometimes limit this pitch to porches while requiring greater slopes for the building proper.

Common types of roofs by name are gable, hip, mansard, gambrel, and flat, figure 17-2.

GABLE ROOF

The gable roof is the most basic sloped roof design. A roof is called a *gable roof* if it has triangular wall shapes beneath the two sloping roof surfaces at each end of the building, figure 17-3. The geometry of a gable roof is shown in figure 17-4.

Fig. 17-1 Roof frame

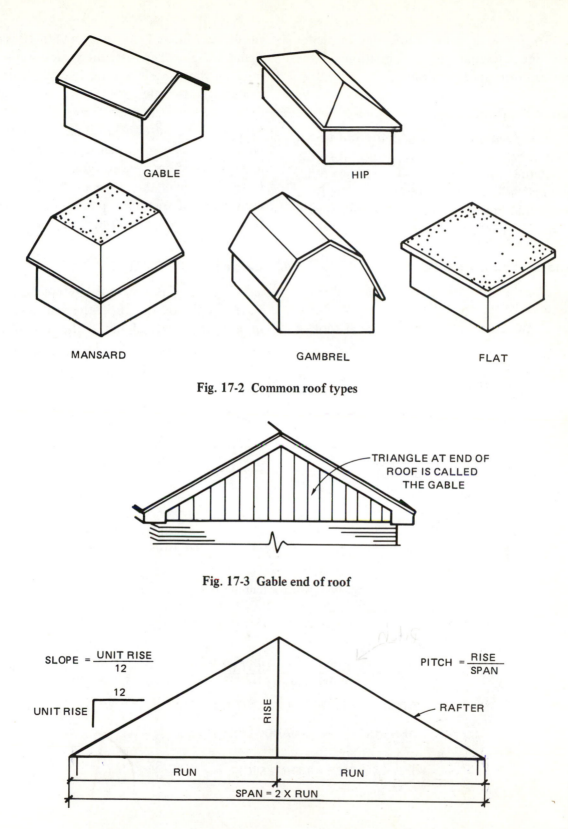

Fig. 17-2 Common roof types

GABLE

HIP

MANSARD

GAMBREL

FLAT

TRIANGLE AT END OF
ROOF IS CALLED
THE GABLE

Fig. 17-3 Gable end of roof

$$SLOPE = \frac{UNIT\ RISE}{12}$$

$$PITCH = \frac{RISE}{SPAN}$$

UNIT RISE

12

RISE

RAFTER

RUN

RUN

SPAN = 2 X RUN

Fig. 17-4 Gable roof geometry

To frame a gable roof, the carpenter must find the following information on the plans or in the specifications:

Span of the Roof

The span of the roof is the out-to-out distance across the building measured to the face of the studs, figure 17-4. It is two times the length of the run of the rafter. The floor plan should give this dimension since the plan was used to frame the exterior walls.

Slope and Pitch of the Roof

Slope of the roof is the rise in inches for one foot of run. For example, a roof that rises at the rate of 4 inches for each foot of

run is considered to have a 4-in-12 slope, figure 17-5. The carpentry definition for *pitch* is the rise divided by the span (pitch = rise/span). Its value is expressed as a fraction reduced to the simplest form (1/8, 1/4, 1/3, 1/2, etc.). Because these values are not usable without adjustment when laying out a rafter, a simpler system is more often used, and that is to determine the slope of the roof. The conversion of fractional pitch data to rise per foot is given in figure 17-6.

Eave and Ridge Details

The *eave* is the roof overhang or edge. Eave details are shown in the wall section or in a special eave detail. The architectural

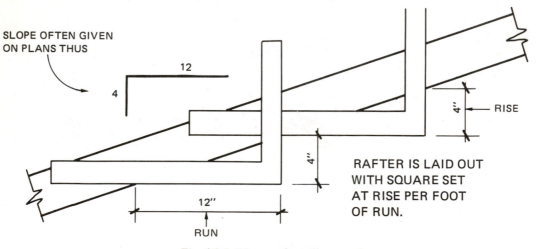

SLOPE OFTEN GIVEN ON PLANS THUS

12

4

4" — RISE

RAFTER IS LAID OUT WITH SQUARE SET AT RISE PER FOOT OF RUN.

12"

RUN

Fig. 17-5 Rise per foot illustrated

Pitch

IF PITCH $= \dfrac{RISE}{SPAN} = \dfrac{1}{3} = \dfrac{UNIT\ RISE}{UNIT\ SPAN}$

AND SPAN = 2 X RUN AND UNIT RUN = 12"

THEN UNIT SPAN = 24 AND PITCH $= \dfrac{1}{3}$ OR $\dfrac{8}{24}$.

WITH UNIT RISE = 8 THEN SLOPE $= \dfrac{8}{12} = \dfrac{RISE/FOOT}{ONE\ FOOT}$

Fig. 17-6 Pitch changed to rise per foot

effect created by the trimwork and the detail of the roof edge is carefully planned. The eave may be closed with all rafters concealed, figure 17-7. It may be open with rafter tails exposed, figure 17-8. The overhang of the roof is dimensioned on these views and is a horizontal distance like the run of the rafter. It is not added to the rafter run in laying out the rafter. Instead, the overhanging portion of the rafter is laid out as a separate step later.

The ridge detail is shown on the section of the building. It may be found on a fireplace chimney section. It may be left up to the carpenter to do as he pleases. Conventional roof framing usually has a ridge board which helps locate and align the rafters, figure 17-9, page 162.

Size, Spacing, and Material of Members

The size of rafters and their spacing is usually given on the section view. The ceiling joists are also identified there. The ceiling joists must be securely anchored to the rafters to hold the joists in place. Otherwise, the side thrust of the rafters would push the walls out and allow the roof to sag, figures 17-7 and 17-8.

Material used in roof framing is properly given in the specifications. It is listed with other framing specs. A typical specification is shown in figure 17-10, page 162.

HIP ROOF

A *hip roof* slopes in four directions instead of two as the gable roof does, figure 17-2.

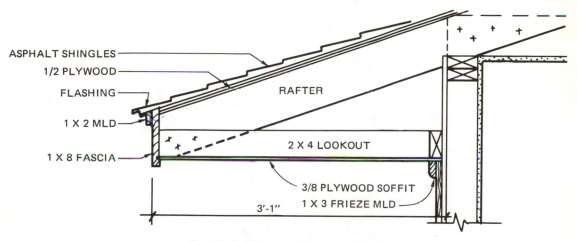

ASPHALT SHINGLES
1/2 PLYWOOD
FLASHING
1 X 2 MLD
1 X 8 FASCIA
RAFTER
2 X 4 LOOKOUT
3/8 PLYWOOD SOFFIT
1 X 3 FRIEZE MLD
3'-1''

Fig. 17-7 Closed cornice eave detail

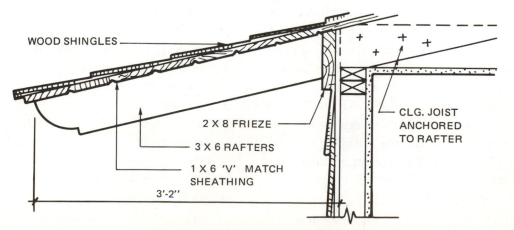

WOOD SHINGLES
2 X 8 FRIEZE
3 X 6 RAFTERS
1 X 6 'V' MATCH SHEATHING
3'-2''
CLG. JOIST ANCHORED TO RAFTER

Fig. 17-8 Open cornice eave detail

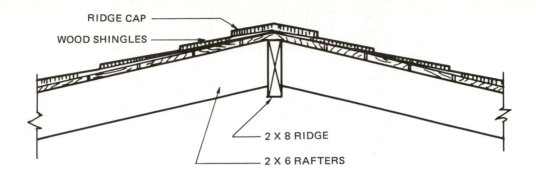

Fig. 17-9 Ridge board detail

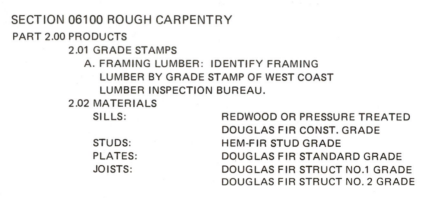

Fig. 17-10 Sample carpentry specs

The slopes are formed exactly as the gable slopes and the same information is required. However, the intersection of the roof planes at the corners of the building requires special layout. The long member at this intersection of the roof planes is called a *hip rafter*. Rafters used in gable roofs and the straight part of hip roofs are called *common rafters*. They run from the exterior wall to the ridge at the required slope. At the corners of the building the common rafters run into the hip rafters before they reach the ridge height. These shortened common rafters are called *jack rafters*. Figure 17-11 shows the different parts of a hip roof.

When viewed from the top, as in a roof plan, the hip rafters always form 45° angles

with the sides, figure 17-11. This means that the jack rafters are in pairs, rights and lefts. It also gives the carpenter the clue to figure the ridge length. By taking half the width of the building off each end of the length of the building, the ridge length is found, figure 17-12. Note that the framing dimensions from the floor plan are used for this calculation.

Sometimes the hips and ridge are indicated on the floor plan by dotted lines. This is a simple way to explain the roof when no roof plan is drawn, figure 17-13. The elevations show the roof to be hipped, but the plan layout confirms it.

MANSARD ROOF

The mansard roof was originated by Francois Mansard (1598–1666), a French

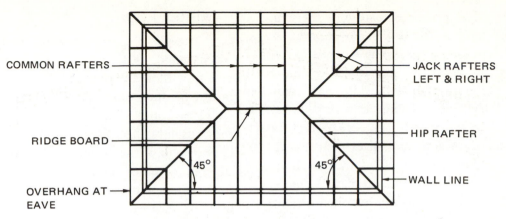

Fig. 17-11 Hip roof framing members

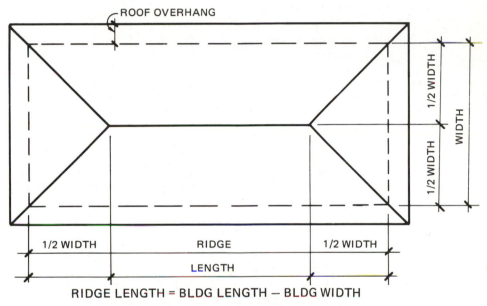

RIDGE LENGTH = BLDG LENGTH − BLDG WIDTH

Fig. 17-12 Roof plan showing ridge length

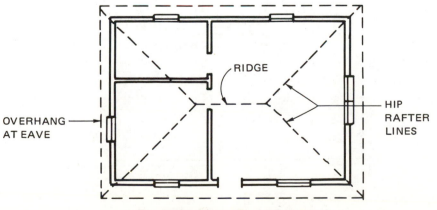

Fig. 17-13 Floor plan with hip roof

architect. The principle feature of the mansard roof style is the very steep slope of the roof. It rises to a break line where the roof becomes flat or nearly so. It also continues around the corners like the hip roof, figure 17-2. For large buildings this sloped roof section becomes merely an edge detail. It lowers the eave line and gives a substantial feeling to the building. This framing is detailed in the section views. It may be framed in wood even on a masonry building. Frames are hung from the main wall providing strong but simple construction for this detail. The flashing details between the mansard section and the main flat roof are important to the success of this roof, figure 17-14. The main advantage of this roof is the additional space provided in the rooms of the upper level.

GAMBREL ROOF

The *gambrel* or *barn roof,* as it is sometimes called, is similar to the mansard. It begins with a steep section but then breaks into a medium slope rather than a flat roof. However, unlike the mansard, the gambrel roof usually has gable ends, figure 17-2.

Framing the gambrel consists of two runs of common rafters at different slopes. In barn construction, in which clear openings are desired for the hayloft, a trussed frame is constructed for each rafter, figure 17-15.

FLAT ROOF

The flat roof, the simplest roof construction, is used extensively in commercial

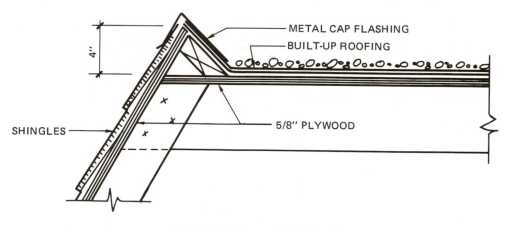

Fig. 17-14 Mansard flashing detail

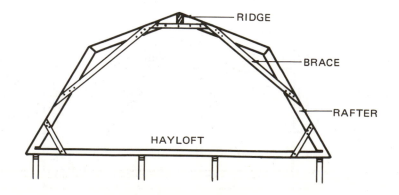

Fig. 17-15 Gambrel trussed frame

construction. Instead of rafters, roof joists are used. The framing is similar to floor framing. Overhangs, girder details, drainage and openings must be studied in the plans. Some roofs are dead flat with joists set level. Others have a slight slope with sloping plates and girders. Small buildings can have tapered joists to give some positive drainage.

ROOF SHEATHING

All wood roofs require sheathing to support the roofing cover. Plywood, structural grade CD, usually is specified over joists and rafters at 24″ o. c. or less. Side joints are reinforced with blocking or ply-clips, figure 17-16. When properly nailed, plywood acts as a diaphragm, or one continuous surface. This adds considerable strength to the assembly for resisting lateral loads of wind and earthquakes.

The nailing schedule usually is given in the specs.

CONTEMPORARY FRAMING METHODS

Conventional framing, or stick building, is the putting up of rafters one at a time. Ceiling joists are set first and braced to the rafters or stiffened in some way later.

Becoming increasingly popular, especially in commercial construction, is the use of prefabricated units. Two such units are trussed rafters and stressed-skin panels.

For greater speed at the construction site, some form of prefabrication is required. The trussed rafter is available from fabricators in many places. It consists of an assembly of two rafters, ceiling joist and bracing, figure 17-17. Overhangs usually are included. They are erected at 24″ o. c. and span from outside

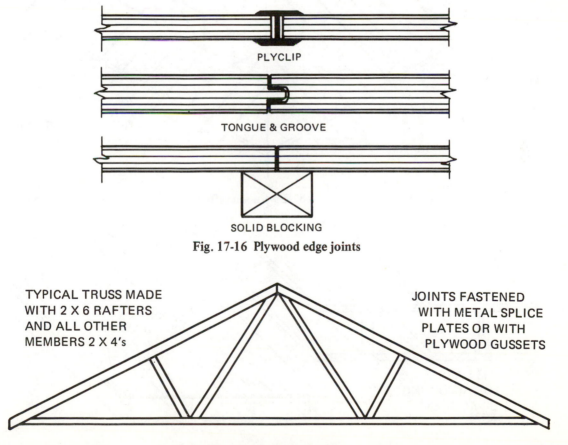

PLYCLIP

TONGUE & GROOVE

SOLID BLOCKING

Fig. 17-16 Plywood edge joints

TYPICAL TRUSS MADE
WITH 2 X 6 RAFTERS
AND ALL OTHER
MEMBERS 2 X 4's

JOINTS FASTENED
WITH METAL SPLICE
PLATES OR WITH
PLYWOOD GUSSETS

Fig. 17-17 Trussed rafter

wall to outside wall. No interior supports are needed because this assembly is a light truss. It is also very efficient in the use of materials. Sheathing is applied to the top surface and the finished ceiling is supported on the lower surface.

Special hip-trussed rafters are available for hip roofs. A series of trusses is built to form the hipped shape. The last few feet are framed conventionally or with special triangular sections, figure 17-18. Trussed rafters are indicated on the floor plan, section views, or in a framing plan. Structural details may be given although specifications for their manufacture are sufficient.

Stressed-skin panels are components composed of joists with one or two surfaces of plywood, figure 17-19. The plywood is nailed and glued to the joists making them act like I-beams. The panels are usually

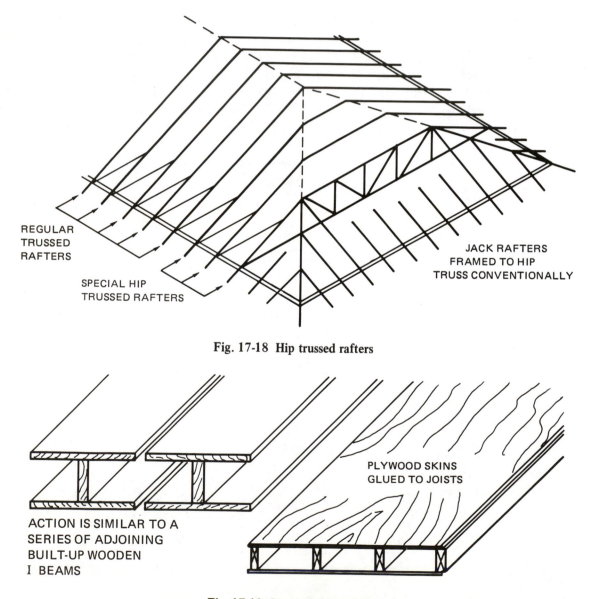

REGULAR TRUSSED RAFTERS

SPECIAL HIP TRUSSED RAFTERS

JACK RAFTERS FRAMED TO HIP TRUSS CONVENTIONALLY

Fig. 17-18 Hip trussed rafters

ACTION IS SIMILAR TO A SERIES OF ADJOINING BUILT-UP WOODEN I BEAMS

PLYWOOD SKINS GLUED TO JOISTS

Fig. 17-19 Stressed-skin panels

four feet wide and a suitable length to span the required space. Greater strength and stiffness are achieved in this manner. Roofs of unusual shape, as well as regular designs, employ stressed skin panels. These components are shop built and delivered to the job for erection. Large units require a crane for easy erection.

SUMMARY

- Roofs are of two general classes: flat or low-sloped, and medium to steep-sloped.

- Flat roofs require built-up roofing.

- Medium to steep roofs are suitable for shingle roofing.

- The gable roof is the most basic design to the other sloped roofs.

- The span of a gable roof is the out-to-out distance across the building.

- The pitch of the roof usually is indicated on the elevations or sections as a slope ratio, rise per foot of run.

- Eave details are planned carefully for architectural effect.

- Ceiling joists tie the rafters together thus keeping them from pushing the walls outward.

- A hip roof has roof slopes in four directions instead of two as in a gable roof.

- Sometimes hip rafters and ridges are indicated by means of dotted lines on the floor plan.

- An advantage of the steep sloped mansard roof is more space in the upper level.

- The gambrel roof is used in barns to provide large hay storage lofts.

- Flat roofs are used extensively for commercial structures.

- Plywood roof sheathing, when properly nailed, serves as a support for the roofing and as a structural diaphragm.

- Trussed rafters and stressed-skin panels are prefabricated roof units developed to speed up the on-site construction process.

REVIEW QUESTIONS

1. What type of roofing is usually used on
   a. sloping roofs?   b. flat roofs?

2. What is the minimum slope for roofs using shingles?
   a. 2 in. per foot   c. 5 in. per foot
   b. 3 in. per foot   d. 8 in. per foot

3. Sketch a gable roof and a hip roof.

4. Which type sloped roof is the most basic to the others?
   a. hip
   b. gambrel
   c. mansard
   d. gable

5. What is the rise per foot of a 1/3 pitch roof?
   a. 3 in.
   b. 4 in.
   c. 8 in.
   d. 12 in.

6. What is the purpose of the ridge board?

7. What drawing gives the size and spacing of the rafters?

8. Why must the ceiling joists be securely anchored to the rafters?

9. Name three kinds of rafters needed to frame a long hip roof.

10. In a roof framing plan of a hip roof, the hip rafters always make an angle of _____ ° with the sides of the roof.
    a. 30°
    b. 90°
    c. 45°
    d. depends upon the pitch

11. What type roof is often used in barn construction to make a large hayloft?

12. Sketch and name three methods used to reinforce side joints in plywood roof sheathing.

13. Trussed rafters are important roof framing components. What is the usual spacing of these units?
    a. 24 in. o. c.   c. 30 in. o. c.
    b. 16 in. o. c.   d. 18 in. o. c.

14. Why do stressed-skin panels have greater strength than conventional framing using the same sized members?

Unit 18
Heavy Timber and Glue-Lam

OBJECTIVES

After studying this unit, the student should be able to:

- tell why heavy timber and glue-laminated (glue-lam) construction is competitive with steel and concrete.
- discuss glue-lam arch construction including types and base connections and ties.

IDENTIFICATIONS

Modern fastenings and engineered systems have enabled heavy timber and glue-lam construction to compete with steel and concrete in new construction. The superior fire resistance of heavy timber construction is also being recognized. When a piece of heavy timber is exposed to fire, the charred exterior of the timber acts as insulation for the uncharred interior of the member. This makes it more fire resistant than steel which loses strength when softened by heat.

Heavy timber construction is a direct descendant of earlier methods used in this country and in Europe. It is a system of framing using heavy floor or roof beams supported on walls or posts. Wood girders also are used to support beams in some buildings. Floor systems of heavy planks are most often used, figure 18-1.

Glue-lam refers to large wooden members that consist of many layers of thin lumber glued together. The length of the member can be made as long as necessary by splicing the layers. Many shapes can be formed by clamping the layers in a jig when gluing, figure 18-2.

Application of glue-lam members is made possible by modern glues and gluing techniques. Seasoning and milling the lumber also are important. Note, however, that only

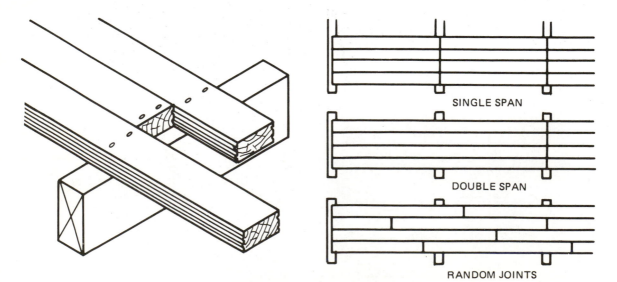

SINGLE SPAN

DOUBLE SPAN

RANDOM JOINTS

Fig. 18-1 Heavy plank floor systems

parallel-to-grain joining of the wood is feasible with gluing techniques, figure 18-3. Glue-lam construction also results in well-seasoned members of any size which is impossible in solid sawn timbers.

GLUE-LAM ARCH CONSTRUCTION

A popular use of glue-lam is the arch. It is especially efficient in framing clear span spaces such as churches, gymnasiums, storage sheds, etc., figure 18-4, page 172. The arches usually rest on a foundation at the floor line of the building. They are spaced 12 to 20 feet apart and form the ribs of the structure. They provide excellent lateral stability for the building. This is a special problem in large open buildings with no cross walls. Enclosing walls fill the spaces between the arches and may be masonry, wood, or metal.

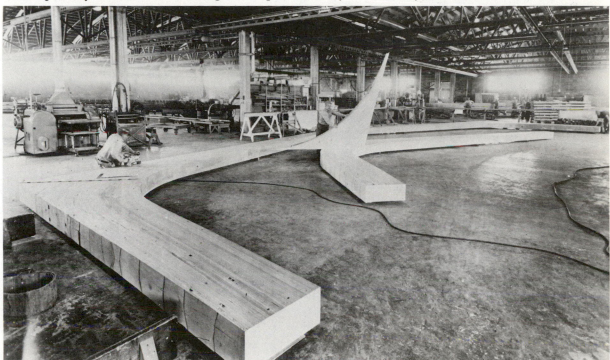

Fig. 18-2 Fabrication of glued laminated arches spanning 81 ft.

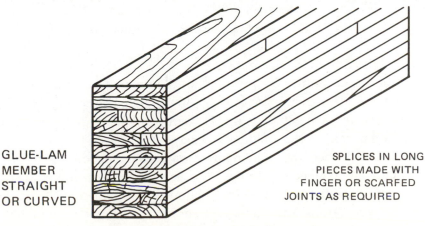

GLUE-LAM
MEMBER
STRAIGHT
OR CURVED

SPLICES IN LONG
PIECES MADE WITH
FINGER OR SCARFED
JOINTS AS REQUIRED

Fig. 18-3 Lumber glued parallel-to-grain

Fig. 18-4 Glu-lam field house arches spanning 203 ft.

Being the major structural element in the building, the location of each arch is shown on the foundation plan. The arches also are indicated on the floor plan. Usually they are left exposed to become part of the architectural design. Sections through the building show the shape and the connecting frame and decking. A detail of the arch is also drawn to show its dimensions and shape, figure 18-5.

Of special importance during the building of the foundation is the tie required at the base of the arch. As ceiling joists tie rafters to prevent movement due to side thrust, so arches have steel ties. The tie usually is encased in concrete just below the floor slab. Sometimes when the thrust is not very large, the concrete floor slab is reinforced to act as the tie for the arches. A foundation detail shows this anchorage and tie arrangement, figure 18-6.

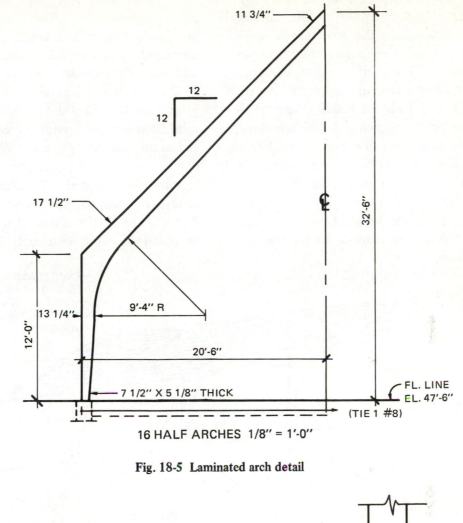

11 3/4"

12
12

17 1/2"

32'-6"

13 1/4"

9'-4" R

12'-0"

20'-6"

7 1/2" X 5 1/8" THICK

FL. LINE
EL. 47'-6"

(TIE 1 #8)

16 HALF ARCHES 1/8" = 1'-0"

Fig. 18-5 Laminated arch detail

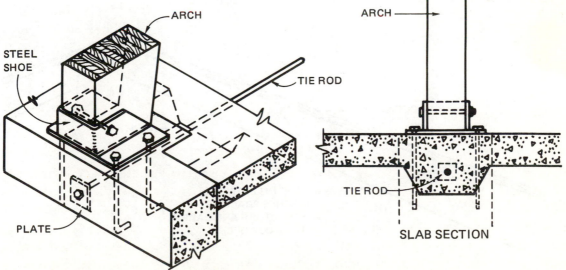

ARCH

STEEL
SHOE

TIE ROD

ARCH

PLATE

TIE ROD

SLAB SECTION

Fig. 18-6 Arch tie in concrete slab

The competitive position of heavy timber construction is due partly to the economy achieved by allowing the structure to show. This is also true of the decking used for floors and for roofs. Glue-lam arches are combined with heavy wood decking that serves as structural deck and finished ceiling, figure 18-7. Depending upon the span, 2-, 3-, and 4-inch decking is used. In some cases wood beams, called *purlins,* span between the arches. The decking is fastened to the purlins, figure 18-8.

Glue-lam arches can be shaped in a variety of ways. Most are designed as three-hinged arches. This allows them to be fabricated in two parts with assembly points at three places, figure 18-9. These assembly points are called hinges and could be pinned connections. A few of the possible three-hinged arch shapes are shown in figure 18-10.

The other design possibility is the two-hinged arch. In this method, the arch is fabricated in one section or else has a more elaborate joint at mid span. When the arch is made in one section, the size of the building is somewhat limited. Circular arches sometimes are built in the two-hinged manner, figure 18-11. The variety of shapes is limited in practical two-hinged arch designs. They

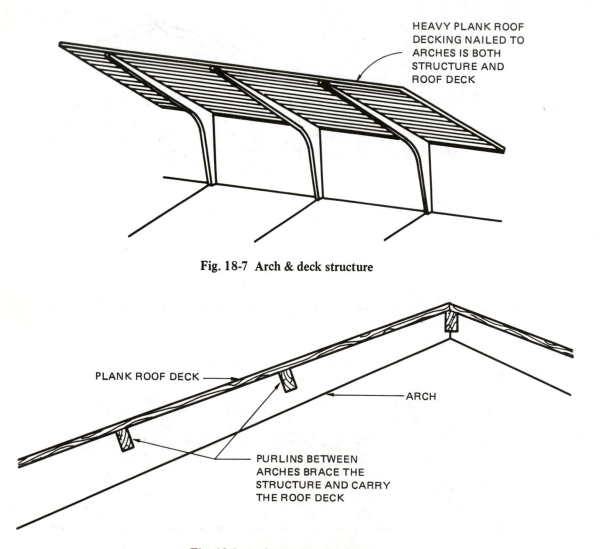

HEAVY PLANK ROOF DECKING NAILED TO ARCHES IS BOTH STRUCTURE AND ROOF DECK

Fig. 18-7 Arch & deck structure

PLANK ROOF DECK

ARCH

PURLINS BETWEEN ARCHES BRACE THE STRUCTURE AND CARRY THE ROOF DECK

Fig. 18-8 Arch & purlins with deck

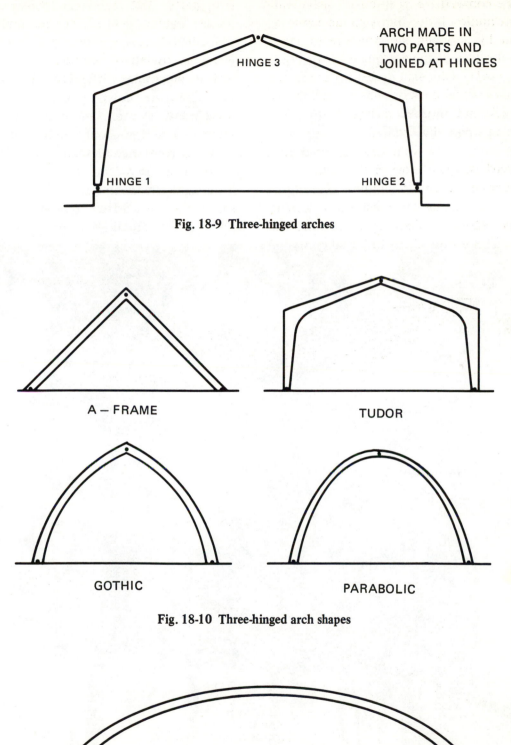

ARCH MADE IN
TWO PARTS AND
JOINED AT HINGES

HINGE 3

HINGE 1   HINGE 2

Fig. 18-9 Three-hinged arches

A – FRAME

TUDOR

GOTHIC

PARABOLIC

Fig. 18-10 Three-hinged arch shapes

HINGE 1   HINGE 2

Fig. 18-11 Circular two-hinged arch

are designed in smooth flowing curves from hinge to hinge because the wood layers must be bent to the shape. Sharp changes in direction are not possible in two-hinged arch construction when made of wood.

GLUE-LAM BEAM AND GIRDER CONSTRUCTION

Straight glue-lam beams are used in a variety of ways in modern heavy timber construction. Commercial buildings, usually of one story, use straight beams for roof support and architectural effect. These buildings are designed around a structural framework that is expressed in the elevations and the interiors, figure 18-12. Connecting these heavy timber elements together in a simple and pleasing way is sometimes difficult. Various connection details are shown in figure 18-13.

Industrial use of heavy timber construction is based not only upon the superior chemical resistance of wood but its economy as well. A very economical framing system using glue-lam beams is the cantilever beam system, figure 18-14. It is used for factory or warehouse buildings.

Fig. 18-12 Architectural use of straight glue-lam roof beams

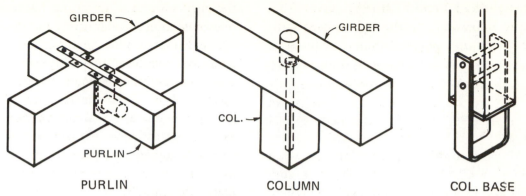

CONCEALED TYPES

Fig. 18-13 Heavy timber connections

Fig. 18-14 Industrial glue-lam roof framing with cantilever girders

MODERN WOOD FASTENERS

Heavy timber construction would have little application today without new methods of fastening. Wood trusses would be very limited in size because the joints between the members are critical. Nails and bolts are quite limited in capacity for these special joints.

Timber connectors for joining wood members consist mainly of split ring connectors and shear plates, figure 18-15. These connectors are much more efficient than bolted connections and make many large wood structures possible. *Split ring connectors* are used in precut grooves between two wood members. The joint is clamped and held together by a bolt through the center of the ring. Multiple rings can be arranged on one bolt, figure 18-16. *Shear plates* are used between metal splice plates and wood members. Bolts are used to hold the parts together. Shear plates especially are useful when the joint needs to be put together and taken apart several times. The connections for glue-lam arches usually have shear plates and metal straps or anchors at peak joint. Methods of indicating split ring connections and shear plates vary, but figure 18-17 gives an example of their use in a truss.

The nailing schedule, as well as splicing of 3- and 4-inch decking when used as structure, is found in the specs. Special nails for side nailing the decking are used. Typical specification notes are given in figure 18-18, page 180.

SPLIT RING

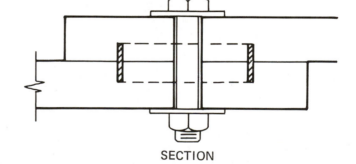

SECTION

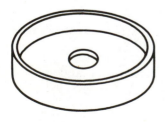

SHEAR PLATE

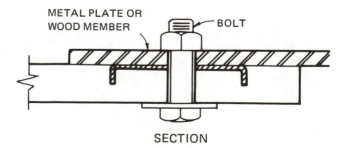

SECTION

Fig. 18-15 Timber connectors

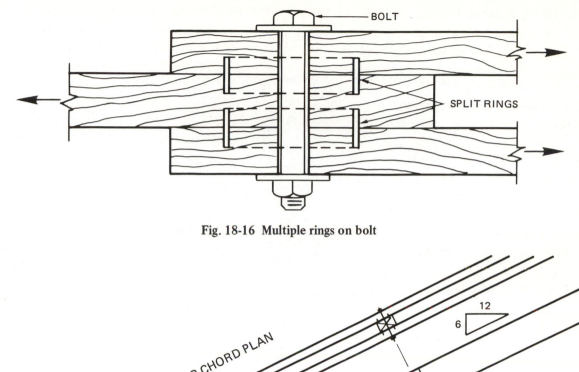

Fig. 18-16 Multiple rings on bolt

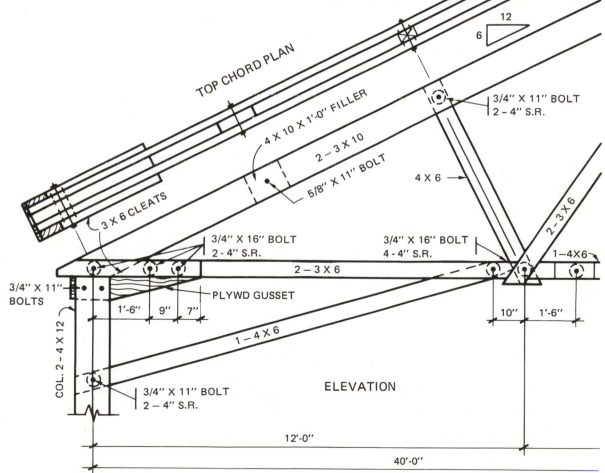

Fig. 18-17 Split-ring truss details

SECTION 06130 HEAVY TIMBER CONSTRUCTION
PART 3.00 ERECTION
    3.01 DECKING
        A. INSTALL DECKING WITH TONGUES UP ON SLOPED ROOFS AND OUTWARD ON FLAT ROOFS.
        B. TOENAIL EACH PIECE TO SUPPORT WITH ONE 40d NAIL AND FACE NAIL WITH ONE 6 IN. SPIKE.
        C. ALL PIECES MUST REST ON AND BE NAILED TO AT LEAST ONE SUPPORT.
        D. COURSES SHALL BE SPIKED TO EACH OTHER AT 30 IN. MAXIMUM SPACING THROUGH PREDRILLED EDGE HOLES.
        E. SPACE JOINTS A MINIMUM DISTANCE OF 4 FT. IN ADJACENT COURSES.

Fig. 18-18 Specifications for nailing deck

SUMMARY

- Modern fastenings and engineered systems have made heavy timber and glue-lam construction competitive systems.

- Heavy timber construction uses heavy wood beams, girders, and posts instead of 2-inch dimension lumber.

- Glue-lam refers to the fabrication of large wood members by gluing together many layers of thin lumber.

- Curved shapes are made by gluing the section in a jig.

- Glue-lam arch construction is an efficient structural system for churches, gymnasiums, storage sheds, etc.

- Arches cause side thrusts at their supports and need to be anchored or tied.

- Glue-lam arch constructed buildings usually have heavy wood decking between the arches.

- Glue-lam arches and decking are usually left exposed.

- Glue-lam arches are designed as two-or three-hinged arches with the three-hinged style being the more versatile.

- Straight glue-lam beams can be fabricated any size and length subject to handling limitations.

- A very economical framing system using glue-lam beams is the cantilever beam system.

- Modern timber connectors, mainly split rings and shear plates, make large wood structures possible.

REVIEW QUESTIONS

1. Give two factors that make heavy timber and glue-lam construction important structural methods today.

2. What kind of floor construction is usually used in the heavy timber system?
   a. concrete slab.   c. heavy plank.
   b. wood joists and deck.   d. steel joists and wood deck.

3. What is the maximum length of a glue-lam member?
   a. 20 ft.   c. 40 ft.
   b. the length of lumber available.   d. the length needed.

4. A laminated arch detail, figure 18-5, gives information about its shape and size.
   a. What is the pitch of the roof this arch forms?

   b. What is the wall height?

   c. What is the width of the building out-to-out on the arches?

   d. How many full arches are used in this building?

   e. What is the size of the arch at its base?

   f. What size tie is used to resist the side thrust of these arches?

5. Arches are designed as two-hinged or three-hinged structures. Illustrate what is meant by these terms.
   a. two-hinged arch
   b. three-hinged arch

6. Name two types of modern connectors.

7. Split ring truss details are given in figure 18-17 for a heavy timber truss.
   a. What size split rings are used?

   b. What size bolts are required with the split rings?

   c. What two sizes of bolts are required without split rings?

   d. What size is the top chord (rafters)?

   e. How wide is this building center to center of columns?

   f. What is the slope of the roof?

   g. What are the maximum and minimum numbers of split rings on one bolt in this detail?

   h. What size is the center section of the bottom chord of this truss?

   i. How thick must the plywood gusset be to fit between the column members?

8. According to the specifications, figure 18-18, heavy decking must be spiked together through predrilled edge holes at a maximum spacing of:
   a. 24 in.   c. 48 in.
   b. 30 in.   d. 60 in.

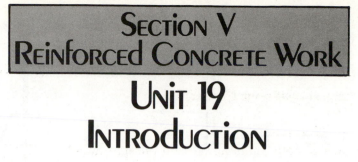

SECTION V
Reinforced Concrete Work
Unit 19
Introduction

OBJECTIVES

After studying this unit, the student should be able to:

- give two important reasons for reinforcing concrete with steel bars.
- explain the numbering system used to designate reinforcing steel bar sizes.
- tell why prestressed concrete was developed and widely accepted.

PRINCIPLES OF REINFORCED CONCRETE

The commercial carpenter must be an expert form builder because of the widespread use of concrete in buildings. Most concrete poured in these forms is reinforced. A basic knowledge of the purpose of reinforcement in concrete helps the commercial builder understand the importance of accuracy in this work.

The location of the steel is calculated carefully by the engineer. If it is not placed properly, the member may fail. Concrete, which is strong in compression, is very weak in tension. Compressive stresses result when a load pushes against a member, figure 19-1. Tensile stresses occur when the load pulls on the member, figure 19-2, page 184. Foundations and columns support loads in compression. There are very few concrete members that support loads in tension alone. However, there are tensile stresses in all beams, suspended floor slabs, retaining walls, and wide footings. All these members have bending stress caused by loads pushing on one side of the member, figure 19-3, page 184. The bottom side of the member is trying to pull apart *(tension)*. The top part

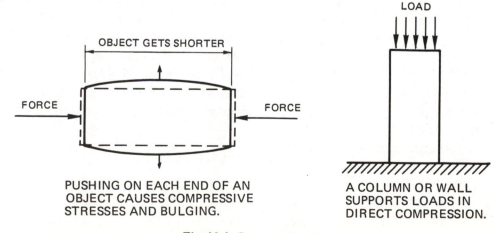

PUSHING ON EACH END OF AN OBJECT CAUSES COMPRESSIVE STRESSES AND BULGING.

A COLUMN OR WALL SUPPORTS LOADS IN DIRECT COMPRESSION.

Fig. 19-1 Compressive stresses

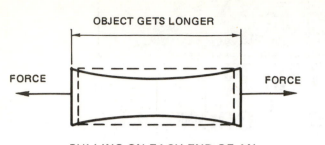

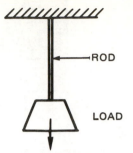

OBJECT GETS LONGER

FORCE ← → FORCE

PULLING ON EACH END OF AN
OBJECT CAUSES TENSILE
STRESSES AND STRETCHING
AND A REDUCTION IN CROSS
SECTION AT ITS CENTER

ROD

LOAD

A ROD OR CABLE
SUPPORTS LOADS IN
TENSION AND TRIES
TO PULL APART.

Fig. 19-2 Tensile stresses

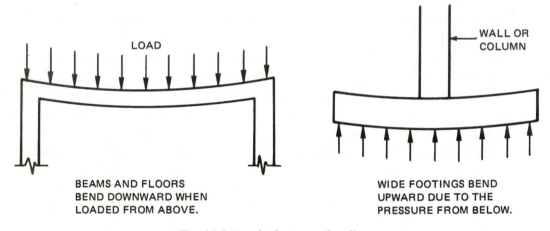

LOAD

BEAMS AND FLOORS
BEND DOWNWARD WHEN
LOADED FROM ABOVE.

WALL OR
COLUMN

WIDE FOOTINGS BEND
UPWARD DUE TO THE
PRESSURE FROM BELOW.

Fig. 19-3 Loads that cause bending

of the member is being pushed together
(compression), figure 19-4. Steel bars are
placed in the concrete near the bottom to
reinforce the weak concrete in the tension
zone, figure 19-5.

More complicated arrangements of reinforcing are required for concrete members
that continue over several supports. The
principle is the same, however, with steel
being placed where the concrete tries to
crack, figure 19-6. The fact that concrete can
be poured any length and is connected
completely at every joint is a structural
advantage. This system is classified as a
continuous structure as if made of one piece.

Another use of steel tensile reinforcement is to resist cracking of the concrete due

to stress caused by shrinkage and temperature
changes. In this case, smaller steel bars are
run at right angles to the main reinforcement.

Although concrete is strong in compression, columns are reinforced because the
steel is also very strong in comparison. The
size of the column can be much smaller if
steel bars are added. Also, many columns in
multi-story concrete buildings have bending
stresses as well as direct compression.

REINFORCEMENT

The kind of material to be used for steel
reinforcement is described in the specs. One
type of reinforcement is reinforcing bars.
Reinforcing bars are round bars with special

184

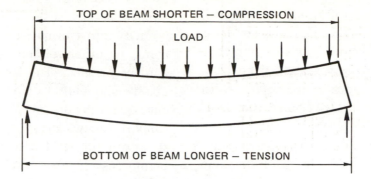

TOP OF BEAM SHORTER — COMPRESSION

LOAD

BOTTOM OF BEAM LONGER — TENSION

Fig. 19-4 Stress in bending member

CONCRETE CRACKS EASILY WHEN PUT IN
TENSION AND MUST BE REINFORCED.

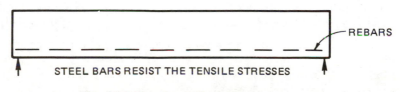

REBARS

STEEL BARS RESIST THE TENSILE STRESSES

Fig. 19-5 Tensile reinforcement

CONTINUOUS BEAM IN TENSION OVER SUPPORT ALSO

SPAN 1   SPAN 2

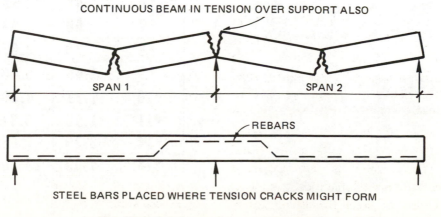

REBARS

STEEL BARS PLACED WHERE TENSION CRACKS MIGHT FORM

Fig. 19-6 Continuous beam reinforcement

lugs or bumps on their surfaces, figure 19-7. These lugs help anchor the steel in the concrete. Sometimes hooks are formed at the end of a bar to give additional anchorage of supports. Standard deformed bars are designated by the number of eighths of an inch in the diameter of the bars. A #4 bar is four eighths or one half inch in diameter. A list of standard bar sizes is given in figure 19-8.

Another type of reinforcement is *welded wire fabric*. It consists of cold-drawn wires welded together to form a mesh. It comes in rolls or sheets. Several different sized wires and mesh patterns are made. It is widely used in floor slabs, especially ground supported slabs. A typical style designation is 4 x 8 – W16 x W10, figure 19-9. This is a new method for designating this reinforcement. The wire size is given as the cross-sectional area rather than the steel wire gauge of the former system. A typical style designation of the previous system is 6 x 6 – 10/10 WWM.

FORMWORK

Of great importance to the widespread use of concrete is the fact that it can be molded in any shape. The wet mixture is poured into forms constructed of wood, metal, or other suitable material in which it hardens or sets. The forms must be of high quality and built to close dimensional tolerances. Formwork must be strong enough to support the weight of the concrete and the stress of pouring. It should also be designed to permit easy removal when the concrete is set.

The specs describe the materials and details of the formwork. They do not show how the forms are to be built. The structural drawings show the layout and sections of the concrete after the forms are removed. The drawings devote much effort to scheduling the reinforcement but provide minimum information about the concrete. Sometimes the architectural details give the other information needed to form the concrete.

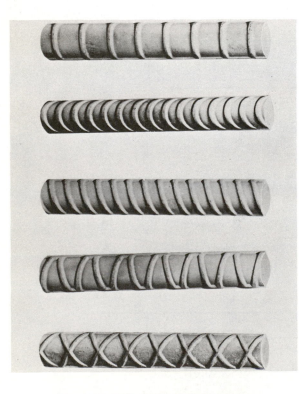

Fig. 19-7 Deformed reinforcing bars

ASTM STANDARD REINFORCING BARS			
BAR SIZE DESIGNATION	AREA* SQ. INCHES	WEIGHT POUNDS PER FT.	DIAMETER* INCHES
#3	.11	.376	.375
#4	.20	.668	.500
#5	.31	1.043	.625
#6	.44	1.502	.750
#7	.60	2.044	.875
#8	.79	2.670	1.000
#9	1.00	3.400	1.128
#10	1.27	4.303	1.270
#11	1.56	5.313	1.410
#14	2.25	7.650	1.693
#18	4.00	13.600	2.257

*Nominal dimensions.

Fig. 19-8 Standard bar sizes

The formwork for a concrete structure constitutes a considerable item of cost even though the formwork is temporary. Particular care should be exercised in its design. Identical units that can be removed and re-used at other locations with a minimum of labor are desirable. The minimum time the forms must remain in place before stripping usually is governed by the local building code and often stated in the specifications. This curing time is a factor in the cost of forming since a long curing time means more forms or else a delayed schedule. Sometimes beams and slabs can be stripped and then braced with posts or shoring. This is called *reposting* or *reshoring* and frees the forms for immediate reuse, figure 19-10.

PRESTRESSED CONCRETE

Prestressed concrete was first used in France but is now becoming increasingly important in the United States. It was developed to make more efficient use of the materials in reinforced concrete. Its greatest application is in members that bend, such as beams, girders, floor slabs, etc. Most precast concrete systems use prestressed concrete (Unit 20, *Precast Concrete Floor Systems*).

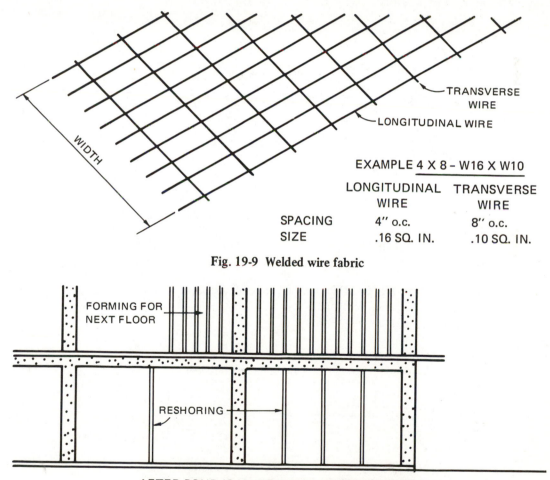

TRANSVERSE WIRE

LONGITUDINAL WIRE

WIDTH

EXAMPLE 4 X 8 – W16 X W10

	LONGITUDINAL WIRE	TRANSVERSE WIRE
SPACING	4" o.c.	8" o.c.
SIZE	.16 SQ. IN.	.10 SQ. IN.

Fig. 19-9 Welded wire fabric

FORMING FOR NEXT FLOOR

RESHORING

AFTER POUR IS MADE AND CURES SUFFICIENTLY, FORMS ARE REMOVED AND A FEW SHORES ARE PUT BACK IN TO BRACE THE STRUCTURE.

Fig. 19-10 Reshoring concrete building

The steel reinforcing in prestressed concrete is stretched, or put in tension, before the concrete is poured. Then when the concrete sets, it bonds to the steel. The steel then is cut off at the ends of the concrete member, but the tension is now resisted by the concrete. The steel holds the concrete together under pressure. This may be compared to holding a row of books together tight enough to lift them as a unit, figure 19-11. The steel is placed toward the bottom of the member, and this tension causes it to buckle upward, figure 19-12. When the load is applied, the member straightens. High strength steel and concrete are used and the concrete is used efficiently.

When the steel is stretched before the concrete is poured, as just described, the process is called *pretensioning* or *bonded prestressing*. This process is used for most precast work where many similar units are made. When the steel is stretched after the member is made, it is known as *post-tensioning*. In these members, tubes are cast in the concrete for the steel to slide through. The steel then is pulled tight by means of hydraulic jacks and anchored at the ends. The concrete must be cured before the steel is stressed. Large girders and multiple members can be post-tensioned at the construction site.

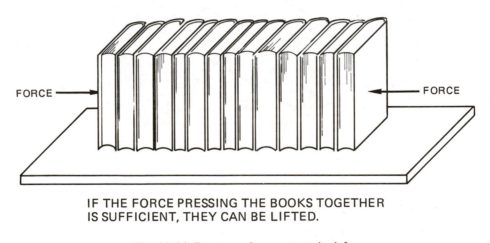

IF THE FORCE PRESSING THE BOOKS TOGETHER
IS SUFFICIENT, THEY CAN BE LIFTED.

Fig. 19-11 Prestressed concrete principle

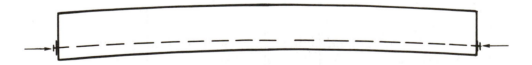

PRESTRESS TENSION IN STEEL AT BOTTOM
OF BEAM CAUSES BEAM TO BUCKLE UPWARDS.
LATER WHEN INSTALLED AND UNDER LOAD,
THE BEAM WILL STRAIGHTEN AND LOOK NORMAL.

Fig. 19-12 Prestress buckles beam

SUMMARY

- The location of steel reinforcement is carefully calculated by the engineer and is critical to the safety of the structure.

- Tensile reinforcement is used where the concrete tends to crack or pull apart.

- Compressive reinforcement is used to help resist loads that try to crush the concrete.

- Continuous structures are possible using reinforced concrete.

- Reinforcing bars have special lugs or bumps to help anchor them in the concrete.

- The size of reinforcing bars is designated by a number representing the diameter in eighths of inches.

- Welded wire fabric, made of cold drawn wires welded together, is widely used in slabs poured on the ground.

- Formwork must be constructed accurately and strongly enough to support the weight of the concrete and the stress of pouring.

- The formwork for a concrete structure is expensive and should be planned for multiple usage where possible.

- Early removal of slab and beam forms is made possible by reshoring.

- Prestressed concrete was developed to make more efficient use of steel and concrete in reinforced concrete structures.

- Prestressed concrete may be pretensioned or post-tensioned.

- Prestressed concrete is becoming increasingly popular.

REVIEW QUESTIONS

1. What is the carpenter's biggest task when working on concrete structures?

2. What can be done to overcome the tendency of concrete to crack when subjected to tensile stresses?

3. Name two parts of a building made of concrete that support load in compression.

4. Where are tensile stresses in a typical concrete building?

5. Choose the points in the following where the concrete will try to crack.

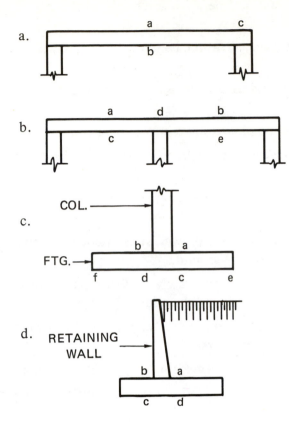

6. Sketch the following:

   a. A steel reinforcing bar.

   b. A piece of welded wire fabric.

7. What is the diameter of a #6 reinforcing bar?

   a. 3/8 in.   c. 3/4 in.

   b. 1/2 in.   d. 1 in.

8. What type of reinforcement is often used in ground supported concrete slabs?

9. What is the procedure called that braces partly cured concrete beams or slabs after the forms have been removed?

10. What effect does prestressing a concrete beam have on its shape before it is loaded? Draw and label a sketch to show the effect.

Unit 20
Floor Systems

OBJECTIVES

After studying this unit, the student should be able to:

- identify two general types of reinforced concrete floor systems.
- explain the difference between one-way and two-way slab systems.
- tell how waffle slabs are formed.

GENERAL TYPES OF REINFORCED CONCRETE FLOOR SYSTEMS

There are a number of reinforced concrete floor systems that have been developed for suspended floor construction. A *suspended floor* is one that spans from one support to another, figure 20-1.

Two general types of reinforced concrete floor systems in common use are cast-in-place and precast. The cast-in-place systems can be classified as "one-way solid slab and beam," "one-way ribbed or joist slab," "two-way solid slab and beam," and "two-way flat plate, flat slab, or waffle slab." The precast systems include "solid slabs," "cored slabs," and "single-tee and double-tee sections." More attention is given here to the cast-in-place systems since the commercial builder must build forms for these.

ONE-WAY SOLID SLAB AND BEAM

One of the most used types of reinforced concrete floor construction is the one-way solid slab and beam. It consists of a solid slab of concrete supported by two parallel beams or walls, figure 20-2. The main reinforcement in the slab runs in one direction only, from beam to beam. This is the reason it is called a one-way slab. The slab is uniform in thickness. The thickness varies from four to six inches depending upon the span and the load. It is used for comparatively short spans of six to twelve feet.

One-way solid slab systems for concrete framed buildings are called *beam-and-girder floors*. The girders span from column to column. Beams are spaced uniformly and frame into the girders at the center, third, or quarter-points. The one-way solid slabs fill in the spaces between the beams. Where the slab spans several beams, it is still a one-way slab but is designed as a continuous element. The formwork is simple and consists of beam forms supported on T-head shores. The slab form is supported on and between the beam forms, figure 20-3.

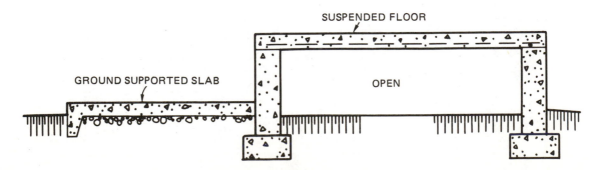

Fig. 20-1 Suspended floor system

One-way as well as two-way solid slabs are scheduled in the structural drawings. Each slab is given a symbol relating it to a schedule which gives the thickness and the reinforcement required. If all slabs are the same thickness, a general note indicates the thickness, figure 20-4.

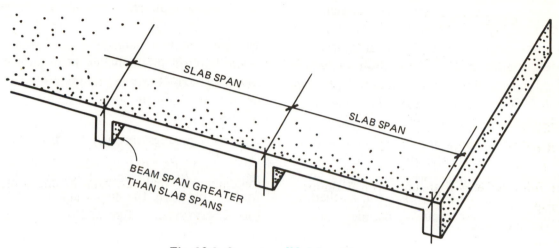

Fig. 20-2 One-way solid slab and beam

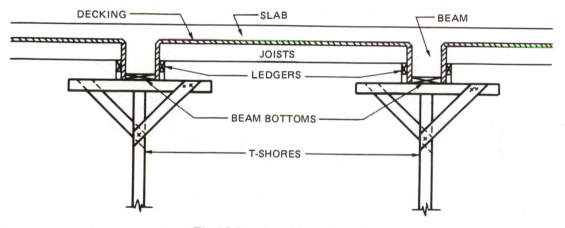

Fig. 20-3 Slab and beam forms

MARK	THICKNESS	MAIN REINFORCEMENT	TEMP. REINF.
1S1	5 1/2	#5 TR @ 18' #5 BOT @ 18	#4 @ 20
1S2	5 1/2	do	do
1S3	5 1/2	do	do
2S1	4 1/2	#4 TR @ 16, #4 BOT @ 16	#3 @ 12
2S2	4 1/2	do	do
RS1	4	#4 TR @ 20, #4 BOT @ 20	#3 @ 20
RS2	4	do	do

Fig. 20-4 Slab schedule

ONE-WAY RIBBED OR JOIST SLAB

One-way solid slabs are too heavy when designed for spans longer than about 12 feet. The dead weight of the concrete is too much compared to the load carried by the slab. By constructing ribs or joists together with a thin slab over them, an economical system is created, figure 20-5. The ribs usually are formed with metal pans that are stripped and reused. The main reinforcement is placed in the bottom of the ribs which are spaced 24 to 35 inches on center. The structural framing plan shows the ribs by dotted lines and dimensions them precisely, figure 20-6. The ribs or joists are also listed in a schedule which gives the reinforcing required.

At times special tapered-end slabs are required, figure 20-7. Special pans must be set to form these ribs. The purpose of the enlarged rib ends is to make the rib larger where it joins the support. This allows the rib to carry more load than it normally could. The pan is held back from the supporting beam two or three inches in order to have room to strip the pan, figure 20-8. This is important since the beam usually is deeper than the joists.

All openings through the floor must be formed and at least one row of bridging is required. *Bridging* consists of cross headers formed between the joists and poured with the floor systems, figure 20-9. Sometimes

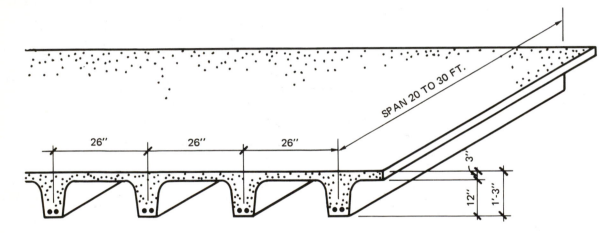

Fig. 20-5 Typical one-way ribbed slab

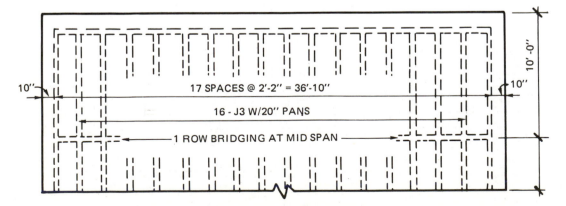

Fig. 20-6 Ribbed slab plan

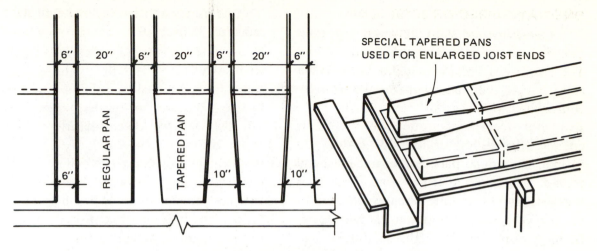

Fig. 20-7 Tapered end forms

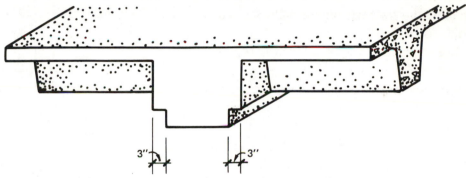

THE METAL PANS THAT FORM THE JOISTS ARE SET 2" TO 3"
OUT FROM THE BEAM TO PROVIDE SPACE FOR THEIR EASY
REMOVAL AFTER THE POUR.

Fig. 20-8 End clearance provided at beams

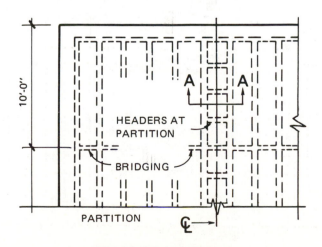

Fig. 20-9 Joist bridging

masonry partitions on the floor occur parallel to the joists and between them. Extra headers are then formed to support the partitions since the thin slab, sometimes 2 1/2 inches thick, is inadequate for this extra concentrated load, figure 20-10.

TWO-WAY SOLID SLAB AND BEAM

When a floor panel is square, or nearly so, the slab is supported on four sides. It is usually more economical to use two sets of main reinforcing bars placed at right angles to each other. The slab is reinforced two ways hence the name, two-way slab. This two-way slab reinforcement is more complicated than the one-way but usually saves on materials, figure 20-11.

Forming a two-way solid slab is simpler than the one-way system. This is because there are fewer beams in the square panel layout. The spans tend to be somewhat longer also. The support beams usually run between columns and are the same size each way for square panels. No girders are needed in this case.

TWO-WAY FLAT PLATE, FLAT SLAB, OR WAFFLE SLAB

These systems are variations of the two-way solid slab. The flat plate system is a solid slab supported by columns but does not have any beams or girders, figure 20-12. The two-way reinforcement brings the loads to the columns. The slab is of uniform thickness which provides a flat ceiling. It is economical to form with no beams or girders. It is used for medium loads and spans such as those found in apartment buildings, office buildings, hospitals, etc.

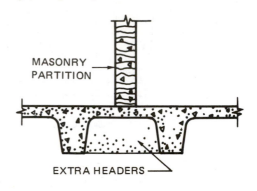

MASONRY PARTITION

EXTRA HEADERS

SECTION A-A

Fig. 20-10 Extra headers

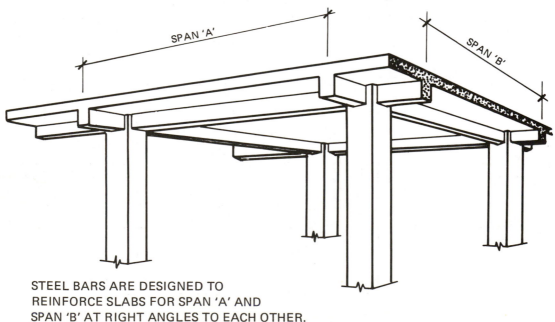

SPAN 'A'

SPAN 'B'

STEEL BARS ARE DESIGNED TO REINFORCE SLABS FOR SPAN 'A' AND SPAN 'B' AT RIGHT ANGLES TO EACH OTHER.

Fig. 20-11 Two-way solid slab and beam

The waffle slab is used for spans and loads that would require extra thick solid slabs. Recesses are formed with removable fillers between ribs running in two directions, figure 20-13. This reduces the dead weight of the system and also adds interest to the exposed ceiling. The fillers, called domes, are available in square sizes ranging from 19 to 63 inches.

The rib widths range from five to eight inches. Forming waffle slabs is done by setting up a smooth level deck upon which the domes are located. The pattern of the domes is found on the structural framing plan. Usually some domes are left out at the columns to provide room for extra concrete to carry the concentration of loads at these points, figure 20-14.

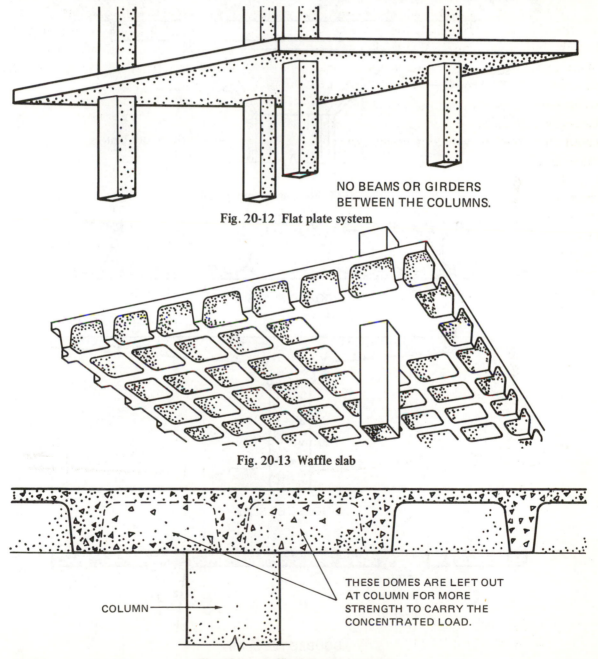

NO BEAMS OR GIRDERS BETWEEN THE COLUMNS.

Fig. 20-12 Flat plate system

Fig. 20-13 Waffle slab

COLUMN →

THESE DOMES ARE LEFT OUT AT COLUMN FOR MORE STRENGTH TO CARRY THE CONCENTRATED LOAD.

Fig. 20-14 Waffle slab column detail

The flat slab system is the oldest and strongest of the two-way systems. It also has no beams or girders framing between the columns. It is suitable for industrial buildings with large loads or other buildings in which the column capitals are not objectionable. The distinguishing feature of the flat slab system is the drop panel and capital at each column, figure 20-15. The drop panel is a square area directly above the column made thicker

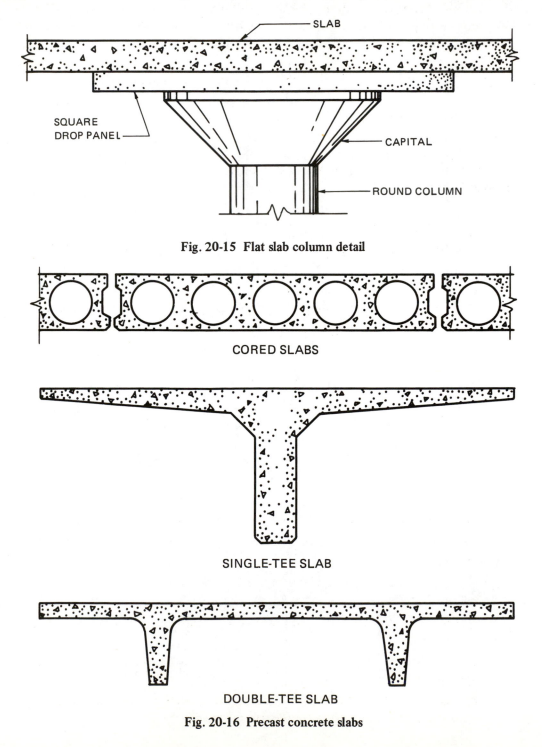

Fig. 20-15 Flat slab column detail

Fig. 20-16 Precast concrete slabs

than the main slab. The column has a flared head or capital which supports the drop panel. The drop panel and column capital increase the capacity of the slab considerably. They prevent the column from punching through the slab because of the heavy loads put on these floors. The forming is still rather simple when compared to the beam and girder system. The structural plans show layout and details quite clearly.

PRECAST CONCRETE FLOOR SYSTEMS

In an effort to have more economically and efficiently constructed floor systems, several precast concrete floor and roof systems are produced. These include solid slabs, cored slabs, single-tee and double-tee sections. Because of the stresses in handling from plant to job site, special steel design is required. For greatest economy, higher strength products are used which yield smaller sections. The method used is called *prestressing* and results in prestressed concrete sections. The sections are detailed by shop drawings for each building. They are erected, anchored, and topped with a concrete topping to provide a smooth surface. Typical sections are shown in figure 20-16, page 198.

SUMMARY

- Suspended concrete floors span from support to support and must be reinforced.
- Slabs on grade may be reinforced for shrinkage and temperature stresses.
- Two general types of reinforced floor systems are cast-in-place and precast.
- One-way slabs have the main reinforcement going in one direction and are rectangular in shape.
- Two-way slabs have the main reinforcement going in two directions and are basically square in shape.
- One-way slabs may be solid or ribbed.
- Two-way slabs may be solid or waffled when supported by walls or beams.
- Two-way slabs supported only by columns are flat plate, flat slab, or waffle slabs.
- Ribbed slabs are formed with removable metal pans.
- Tapered ends on the ribs of one-way ribbed slabs require special forms.
- Cross headers are required between ribs to carry masonry walls when the walls run parallel to the joists.
- Flat plate slabs are solid slabs supported by columns with no beams or column capitals.
- Waffle slabs are formed with removable square domes arranged on a smooth deck.
- The flat slab system has no beams but supports the slab by means of columns with tapered capitals and drop panels.
- Precast concrete slab sections, usually prestressed, are manufactured in a variety of sections for fast construction in the field.

REVIEW QUESTIONS

1. Sketch the following:

   a. ground supported slab.

   b. suspended slab.

2. Name two general types of reinforced concrete floor systems. Which type is the carpenter more interested in?

3. What kind of floor slabs are used in beam-and-girder systems?

4. Sketch a T-shore as used to support beam forms.

5. What is done to make long one-way slabs lighter and more economical?

6. Why are tapered ends sometimes required in ribbed slabs?

7. What is the name given to a slab that is square and supported on four sides?

8. When two-way reinforced slabs are supported only by columns, what are they called?

9. What two-way slab system is the oldest and strongest and has no girders or beams?

10. How are waffle slabs formed?

11. Sketch a flat slab column with capital and drop panel.

12. Sketch the following precast concrete floor sections.
    a. cored slab.

    b. double-tee.

Unit 21
Walls and Columns

OBJECTIVES

After studying this unit, the student should be able to:

- explain the purpose of shear walls in buildings.
- sketch the shape of a cantilever retaining wall.
- tell how columns are identified on structural plans.

SECURITY WALLS

Above grade concrete walls are similar to foundation walls (Unit 13) but not as common. They are used when great strength or security is required. Security here means resistance to fire or to break-in. Bank vaults and safety deposit rooms usually are enclosed with thick heavily reinforced concrete walls. Solid concrete walls also are good sound barriers and might be used between apartments or around soundproofed testing laboratories.

SHEAR WALLS

The superior strength of reinforced concrete makes it good material for shear walls in buildings. A *shear wall* is used to brace a tall framed building against wracking and sidesway, figure 21-1. An example of a shear wall is the elevator and utility core of a tall building which is constructed of poured concrete, figure 21-2. The core serves as structural support for the floors and encloses the elevators in a fireproof shaft. Emergency stairs also may be included in the core.

TILT-UP WALLS

Reinforced concrete walls also are used for the outside walls of a building. They usually are not poured in place. The *tilt-up method* is used, figure 21-3. This method consists of precasting the wall panels at the construction site. Forms are set, reinforcement

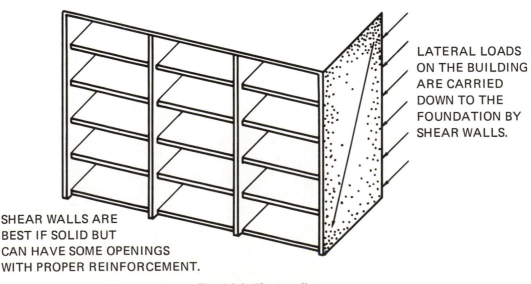

LATERAL LOADS ON THE BUILDING ARE CARRIED DOWN TO THE FOUNDATION BY SHEAR WALLS.

SHEAR WALLS ARE BEST IF SOLID BUT CAN HAVE SOME OPENINGS WITH PROPER REINFORCEMENT.

Fig. 21-1 Shear walls

placed, and the wall poured. To assist in lifting the panel and putting it in place, inserts are provided. Careful placement of these lifting inserts is vital in preventing the walls from cracking when moved. The wall sections are anchored to the foundations and to each other by bolting or welding. Weld plates or clips are cast in the concrete at appropriate places. Since these concrete walls are constructed at the job site, the construction drawings usually provide complete instructions. Shop drawings are not needed for these

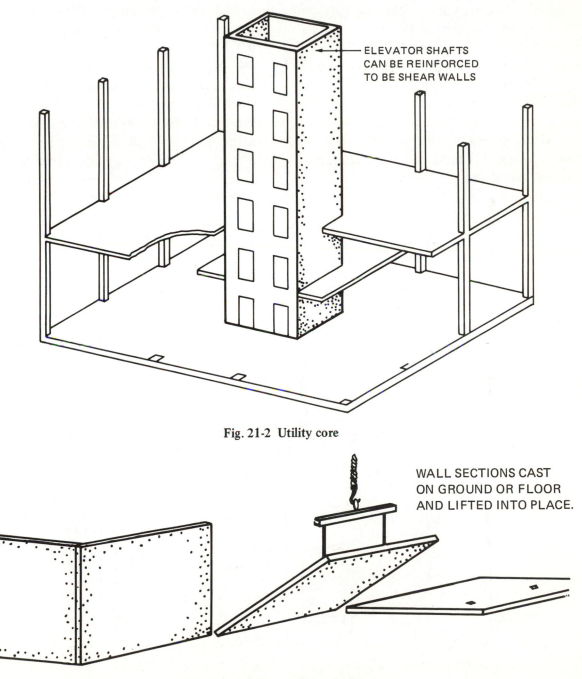

ELEVATOR SHAFTS CAN BE REINFORCED TO BE SHEAR WALLS

Fig. 21-2 Utility core

WALL SECTIONS CAST ON GROUND OR FLOOR AND LIFTED INTO PLACE.

Fig. 21-3 Tilt-up wall construction

panels. However, careful reading of the specs is necessary.

RETAINING WALLS

A *retaining wall* is a free-standing wall whose purpose is to hold back earth or other material. Basement walls hold back earth but they have the building resting on them. This helps hold them in place, but reinforcing is necessary to keep them from bending because of the earth pressure, figure 21-4. Note that the bars are placed on the tension side of the wall.

True retaining walls are of two types, gravity walls and cantilever walls, figure 21-5. Gravity walls must be very heavy to keep from tipping over. They are constructed of brick, stone, or concrete. Gravity walls are usually used for low walls 5-6 feet or less in height. When a wall is high, much material is needed so the cantilever wall is used. It is made of reinforced concrete and has some interesting features.

The cantilever wall has a wide footing, part of which extends under the earth being supported. The weight of the earth behind the wall helps keep the wall from turning over, figure 21-6. The upright part of the wall, called the stem, usually is tapered. This complicates forming the wall but saves concrete and is strong enough. The stem must be reinforced so it will not crack either at the footing or anywhere up the wall. Steel bars are again placed in the tension zone of the wall where it tries to crack, figure 21-6.

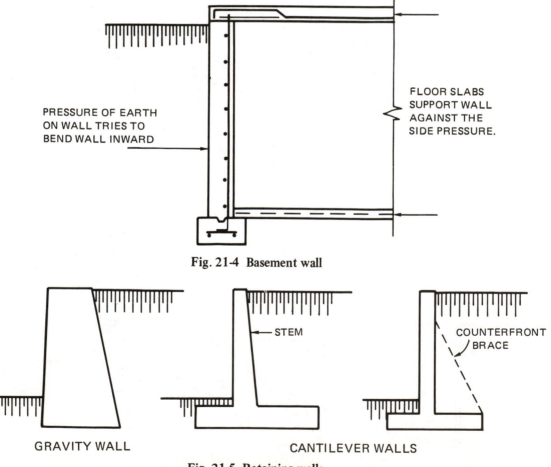

Fig. 21-4 Basement wall

Fig. 21-5 Retaining walls

The footing for retaining walls is poured first with the footing steel in place. In addition, dowels are set carefully in a line for the wall reinforcing to be tied to later, figure 21-7. The wall forms are set after the footing is completed. These walls are detailed carefully on the structural drawings. Provision for weep holes, backfill, and finishing also are included. The necessary quality of materials is given in the specifications.

COLUMNS

Reinforced concrete columns are classified according to shape and compatible reinforcing pattern. Square and rectangular columns are classed as *tied* columns. Both tied columns and spiral columns are commonly used in building construction, figure 21-8, page 206.

Reinforcement in columns is compressive reinforcement for the most part. The steel shares the load with the concrete. Vertical bars, at least four for tied columns and six for spiral columns, are spaced evenly around the column. Since the bars are quite slender, they try to buckle under load. Ties consisting of small bars bent to the shape of the column are used to contain the steel, figure 21-9a, page 206. In round columns, a continuous spiral is used to hold the steel

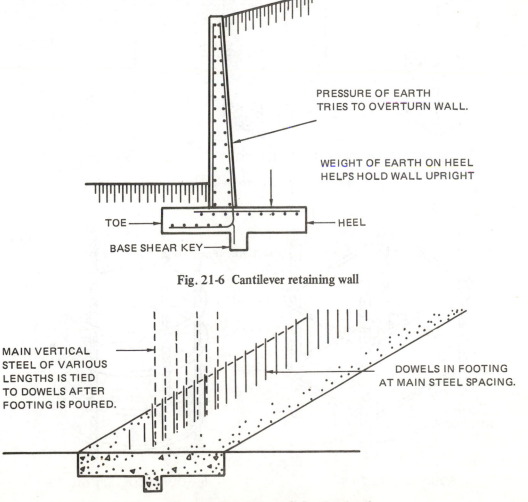

PRESSURE OF EARTH
TRIES TO OVERTURN WALL.

WEIGHT OF EARTH ON HEEL
HELPS HOLD WALL UPRIGHT

TOE

HEEL

BASE SHEAR KEY

Fig. 21-6 Cantilever retaining wall

MAIN VERTICAL
STEEL OF VARIOUS
LENGTHS IS TIED
TO DOWELS AFTER
FOOTING IS POURED.

DOWELS IN FOOTING
AT MAIN STEEL SPACING.

Fig. 21-7 Retaining wall footing detail

in the column, figure 21-9b. The number and size of the column bars can vary within design limits. This enables the engineer to vary the strength of the column without changing its size. In multi-story buildings, the loads accumulate from the top to the bottom. Therefore, the largest columns are at the bottom. By varying the steel, sometimes two or three stories can have the same column size. This simplifies the formwork and reduces the cost of the structure.

The columns are identified on the framing plans by the structural grid coordinates,

figure 21-10. The shape of the column is shown, but its size is given only by scale. To form the columns, more information must be found. The column schedule, especially prepared to give the reinforcing details, also gives data for forming. The size of the column at each floor level is given. The elevation of each floor and the floor to floor distance is also shown, figure 21-11.

Column splice details help to show the transition from one column to the next, figure 21-12, page 208. The steel from the lower column extends above the next floor.

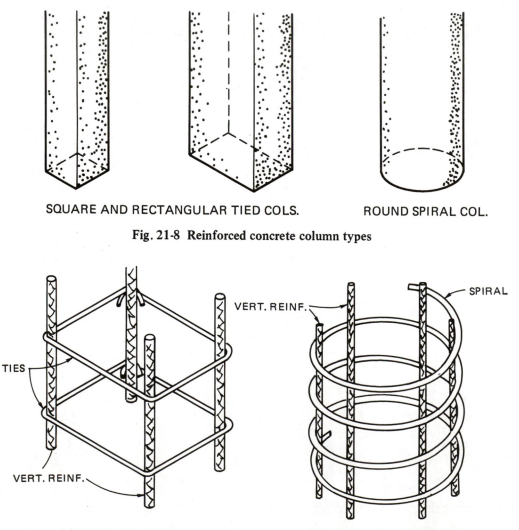

SQUARE AND RECTANGULAR TIED COLS.   ROUND SPIRAL COL.

Fig. 21-8 Reinforced concrete column types

a. SQUARE COL. TIES   b. ROUND COL. SPIRAL TIE

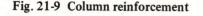

Fig. 21-9 Column reinforcement

After that floor is poured, the next higher column is formed with its steel tied to the projecting bars. Normally the columns are poured several hours or even the day before the next slab pour. This allows the tall column of fresh concrete to settle and shrink before the slab pour is made. It also steadies the formwork and helps in the final adjustment of the slab forms.

Many other shapes and details are used for concrete columns in buildings because

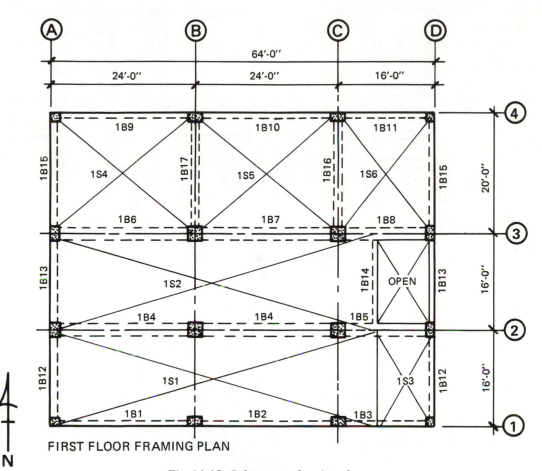

FIRST FLOOR FRAMING PLAN

N

Fig. 21-10 Columns on framing plan

COLUMN MARK	B2, B3 C2, C3	A2, A3 D2, D 3	B4, C4	A1, A4, B1 C1, D1, D4	
ROOF					
COL. SIZE	24 X 24	18 X 24	18 X 20	18 X 18	
VERT. BARS	8 – #8	8 – #7	4 – #9	4 – #9	
1 ST FL. TIES	#4 @ 24	#3 @ 18	#3 @ 18	#3 @ 18	
COL. SIZE	24 X 24	24 X 24	24 X 24	18 X 18	
VERT. BARS	12 – #9	8 – #8	6 – #9	6 – #9	
BSMT TIES	#4 @ 24	#4 @ 24	#4 @ 24	#3 @ 18	

Fig. 21-11 Column schedule

concrete is a plastic material. After the cover of concrete required for fireproofing the steel has been provided, any shape or texture the designer wants is possible. Some special column shapes are shown in figure 21-13.

Since architect's designs sometimes are unusual in their structural usage of reinforced concrete, more sections and details are required to give exact profiles, but the principles are much the same.

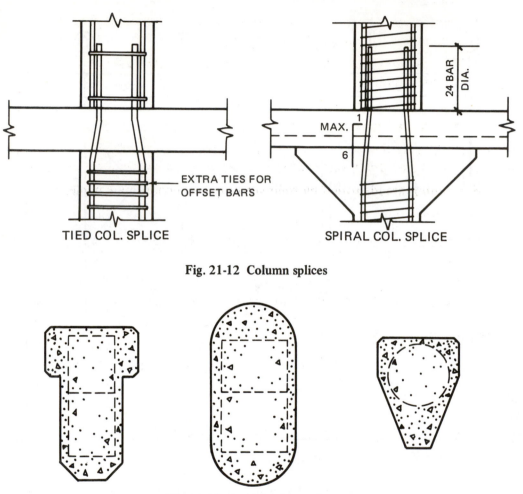

Fig. 21-12 Column splices

Fig. 21-13 Special column shapes

SUMMARY

- Reinforced concrete walls are used when security from fire or break-in is needed.

- Shear walls in tall buildings are sometimes made of reinforced concrete.

- Shear walls brace the building against wracking and sidesway due to wind and earthquake forces.

- A method of casting wall sections lying down and later lifting them into place is called tilt-up construction.

- A retaining wall is a free-standing wall whose purpose is to hold back earth backfilled against it.

- True retaining walls are of two types, gravity walls and cantilever walls.

- Cantilever retaining walls are built of reinforced concrete.

- Cantilever retaining walls consist of a tapered vertical section anchored to a wide footing which extends under the earth the weight of which helps keep the wall from overturning.

- Reinforced concrete columns are divided into two classes, square-tied columns and round spirally reinforced columns.

- The reinforcement in concrete columns helps carry the load in compression.

- The tendency for the steel bars to buckle is resisted by ties arranged around the bars at close intervals.

- By varying the amount of steel at a section, the capacity of the column can be varied without changing the concrete dimensions or the forms.

- Columns are identified by their structural grid coordinate name.

- Column sizes and reinforcement are given in a column schedule.

- Columns are poured sometimes ahead of slabs to allow the concrete to settle and to help stabilize the forms.

- After the covering of concrete required for fireproofing the steel has been provided, any shape or texture the designer wants is possible.

REVIEW QUESTIONS

1. Name two places inside a commercial building where poured concrete walls might be used.

2. What is the purpose of shear walls in a building?

3. Sometimes a tall building is designed around a reinforced concrete core. Which of the following functions are realized by this core?
   a. enclosing the elevators.   c. acting as shear walls.
   b. helping support the floors.   d. enclosing emergency stairs.

4. Walls poured horizontally at the site and later put in place with a crane are called:
   a. precast walls.   c. tilt-up walls.
   b. poured-in-place walls.   d. retaining walls.

5. Sketch the following:
   a. a gravity retaining wall.

   b. a cantilever retaining wall.

6. How is a basement wall reinforced to resist the earth pressure against it?

7. Why is the stem of a reinforced concrete retaining wall usually tapered even though this makes the formwork more difficult?

8. What is the purpose for setting dowels in the footing of a retaining wall?

9. The strength of a reinforced concrete column can be varied by changing the amount of steel without changing its size. Why is this done in multi-story construction where the loads vary from floor to floor?

10. Where is the size, reinforcing, and floor to floor distance for reinforced concrete columns found?

11. Name the two types of reinforced concrete columns used in buildings.

12. Why are concrete columns poured several hours before the beams and slabs which they support?

Unit 22
Special Construction

OBJECTIVES

After studying this unit, the student should be able to:

- explain how stairs are indicated on floor plans and details.

- give the purposes for ramps.

- tell why concrete shell structures are especially strong.

STAIRWAYS AND RAMPS

Reinforced concrete stairways are used in most concrete framed buildings. For convenience, they are formed and poured after the floors are completed. Stairs are essentially a very complicated structure. The floor plan indicates the location of the stairs and the direction of travel, figure 22-1. When the stairs rise through the cutting plane used to draw the floor plan, they are cut off with a break line. Stairs usually have landings at mid-height of the story. A landing below the down flight shows in the floor plan also. The different floor plans should be checked carefully to be sure the stairs are drawn correctly.

The stair sections, architectural and structural, should be studied next. Dimensions for forming and reinforcing the stairs are shown in the structural sections, figure 22-2, page 212. Because the stairs are poured after the floors, provision must be made for the joints between the parts. Notches or ledges are formed to receive the stair sections. Short reinforcing bars, dowels, or bond bars are cast in the floor or beam at the notch. These bars tie the stairs to the supporting member.

The number of risers (the vertical part of the stair step) usually is given with the dimension of each and the total height of the flight. Careful measurement of the actual floor-to-floor distance should be made. Making all the risers equal in a flight is required by building codes. Uneven risers are a safety hazard. The drawings may dimension the height of the flight and let the builder be responsible for dividing it into equal risers, figure 22-2.

The footing for the stairs from the basement floor is a pad cast into the floor slab, figure 22-3, page 212. Dowels should

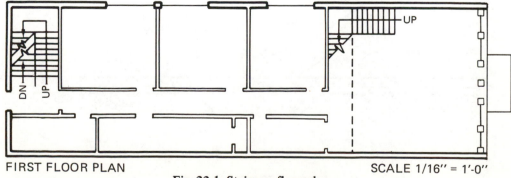

FIRST FLOOR PLAN

SCALE 1/16" = 1'-0"

Fig. 22-1 Stairs on floor plan

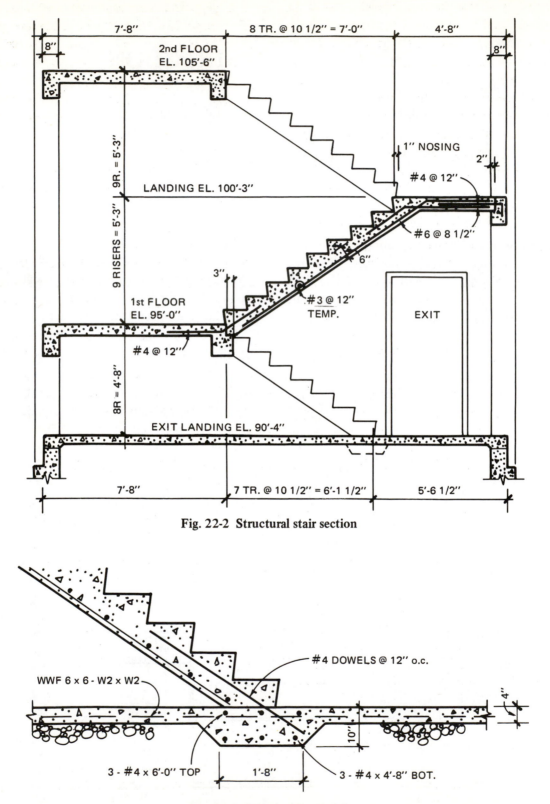

Fig. 22-2 Structural stair section

Fig. 22-3 Stair footing in slab

212

extend out of this footing into the stairs. An alternate detail is to place a footing pad below the slab, figure 22-4. A slab on fill could be poured later in this case.

An enlarged detail of the step shows the angle of the riser and the radius desired at the nosing (the overhanging rounded edge of a stair tread). The tread distance is measured to the base of the angle and does not include the nosing, figure 22-5. Some steps may have special nosing inserts for safer, longer wearing treads, figure 22-6, page 214.

Ramps are used for gradual changes in height for foot traffic, disabled people in wheel chairs, and for automobiles. They are essentially sloping slabs and require careful layout when formed. The overall rise and run

of a ramp is dimensioned on a section, figure 22-7, page 214. Like stairs, ramps are poured after main floors and require reinforced construction joints. If the ramp is between two supporting walls, a slot is formed to support the ramp slab. Later when the ramp is formed, the slot locates it exactly, figure 22-8, page 214.

THIN-SHELL ROOF SYSTEMS

Because of the plastic nature of concrete, it can be molded into almost any shape. Cast-in-place, thin-shell concrete is an interesting structural solution. These structures derive their strength mainly from the curvature of the surfaces. Often there is a double curvature, which is especially strong for the

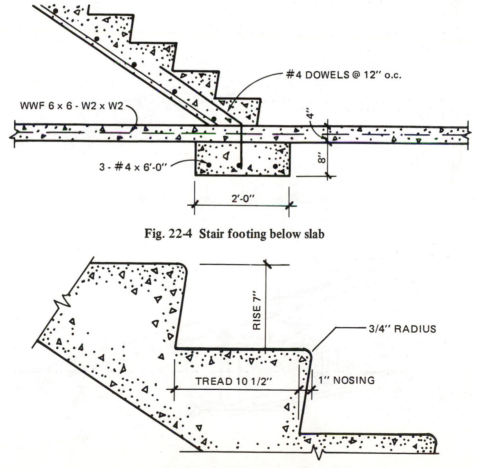

#4 DOWELS @ 12" o.c.

WWF 6 x 6 - W2 x W2

3 - #4 x 6'-0"

4"

8"

2'-0"

Fig. 22-4 Stair footing below slab

RISE 7"

3/4" RADIUS

TREAD 10 1/2"

1" NOSING

Fig. 22-5 Enlarged step detail

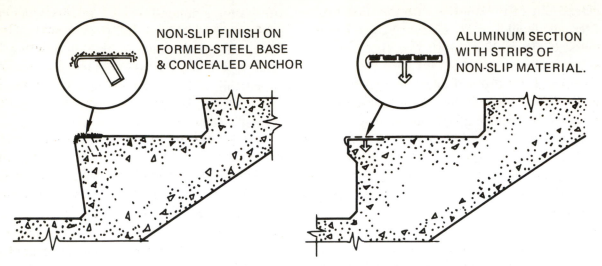

Fig. 22-6 Safety nosings for concrete steps

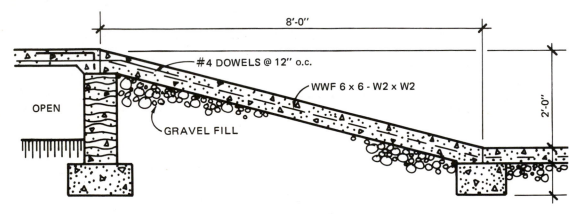

Fig. 22-7 Section of a concrete ramp

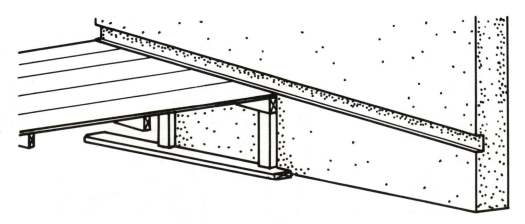

Fig. 22-8 Ramp formed to slot in wall

amount of material used. The egg is nature's best example of this principle. The construction of roof systems of thin-shell concrete is partly limited by the cost of forming. Reuse of form sections, dividing the structure into parts, and using selected geometric shapes have made them practical.

The dome, a traditional form, is one successful shape. A section through the center line gives the curvature and section properties, figure 22-9. Since most of the stresses in a dome are compressive and well distributed, rather thin sections are possible. Reinforcement is used to distribute local concentrations to the area around it. A tension ring must be provided at the base, or else a buttress type foundation must be used,

figure 22-10. It should be emphasized that forming is expensive for large domes. One interesting, although not typical, solution is to build a mound of earth shaped to the inside dimensions of the dome. The dome is poured and the earth excavated to form the building space.

Another thin-shell system for buildings that have columns is the *hyperbolic paraboloid.* This is a double curvature shape that can be used in square or rectangular bays to cover large areas, figure 22-11, page 216. It is also used with twin supports for dramatic roof construction. The forming of these double curvature shapes is not as complicated as it would seem. All these shell planes are formed with straight members. They are set on

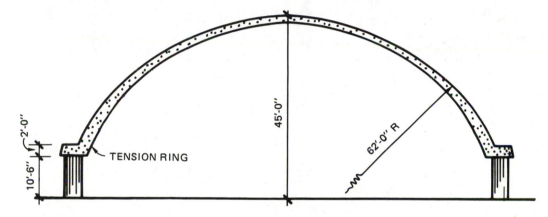

Fig. 22-9 Section of spherical concrete dome

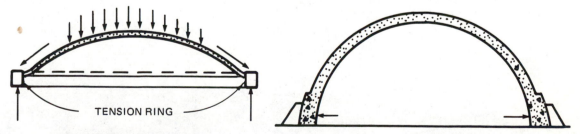

LOADS ON THE DOME ARE CARRIED DOWN TO THE BASE. THE OUTWARD THRUST IS RESISTED BY THE TENSION RING ACTING LIKE A HOOP.

SIDE THRUST OF DOME MAY BE RESISTED BY HEAVY SECTIONS AND BUTTRESSES AT BASE.

Fig. 22-10 Tension ring and buttress

supports that cause the form to be generated, figure 22-12. A covering is then placed over the straight members to blend the curves. The concrete is placed on this form.

Arches and vaults are also built of reinforced concrete. Ribbed arch construction has been used for large clear span structures, figure 22-13. Elliptical and parabolic shells are also popular, figure 22-14.

These special shapes in reinforced concrete are limited only by the architect's imagination and the builder's skill to build them. The commercial carpenter who specializes in formwork should not be surprised to find some of these forms in modern structures. Careful study of the drawings leads to successful construction of these dramatic structures.

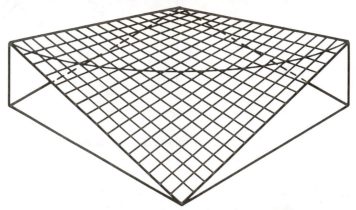

Fig. 22-11 Hyperbolic paraboloid roof structure

Fig. 22-12 Straight lines form hypar shell

Fig. 22-13 Ribbed-arch concrete shell used in hangar construction

Fig. 22-14 Parabolic roof used for a church design

SUMMARY

- Reinforced concrete stairs are built after floor slabs are in place.

- Floor plans indicate the direction of travel, number of riser, and required landings.

- Dimensions for forming and reinforcing details are shown in the structural sections.

- All risers in any one flight of stairs should be the same for safety.

- Ramps are used for gradual changes in height for foot traffic and wheeled vehicles.

- Thin-shell structures derive their strength mainly from their curvatures.

- The dome is a traditional shell form requiring a tension ring about its circular base.

- The hyperbolic paraboloid is a double curvature shell that can be formed on a framework of straight members.

- Many special shapes in concrete are possible when there are imaginative architects to design them and skilled builders to construct them.

REVIEW QUESTIONS

1. Stairways in most concrete framed buildings are constructed of:

   a. precast concrete.

   b. steel with concrete filled treads.

   c. poured-in-place reinforced concrete.

   d. pressure treated wood.

2. When are concrete stairways constructed as related to the main structure?

3. How is the direction of travel, whether up or down, indicated on the floor plans?

4. Sketch the section profile of a typical poured concrete step. Indicate the tread distance and the nosing.

5. What can be done to concrete steps to make them safer and longer wearing?

6. Give two common uses for ramps in buildings.

7. What is nature's best example of a strong double-curvatured, efficient structure?

8. What item of expense in building thin-shell structures often limits their use?

9. A dome carries its loads largely in compression which results in continuous side thrusts at the base of the dome. Give two ways used to resist the side thrusts at the base of a dome.

10. Why are so many unusual shapes possible in reinforced concrete construction?

11. Answer the following questions based on figure 22-2, Structural Stair Section.

a. What is the total number of risers in this set of stairs?

b. What is the dimension of one riser?

c. Are all the risers the same in this set of stairs?

d. What is the tread dimension?

e. How far out does the nosing extend beyond the tread dimension?

f. What reinforcement is required to connect stairs to first floor slab?

UNIT 23
MATERIALS AND LAYOUT

OBJECTIVES

After studying this unit, the student should be able to:

- give typical sizes of brick and concrete masonry units.
- sketch three bond patterns used in modern masonry work.
- tell how the difference between actual and nominal sizes of masonry units affects layouts.

BRICK

Bricks are solid rectangular units formed of clay and shale and burned in a kiln. Most modern bricks now have holes punched in them to conserve material, reduce the weight, and improve the bond. Solid bricks are available for special purposes such as for floors and special details, figure 23-1. Most modern bricks are produced in modular sizes. That is, the brick size is designed so that the dimensions of the brick plus the size of the mortar joint will equal a multiple of the 4-inch grid system. A number of modular units is available. The units are listed in figure 23-2.

Two of the many types of bricks are common or building bricks and face bricks. *Common bricks* are made from ordinary clay or shale. They do not meet special standards of color, design or texture. Sometimes overruns of special bricks are sold as common bricks to dispose of them. Common bricks are less expensive than face bricks and usually are used for concealed work.

Face bricks are produced to meet special standards for color, texture, absorption, uniformity, and strength. More care is exercised in the selection of materials and in the burning of these bricks. The exposed surfaces

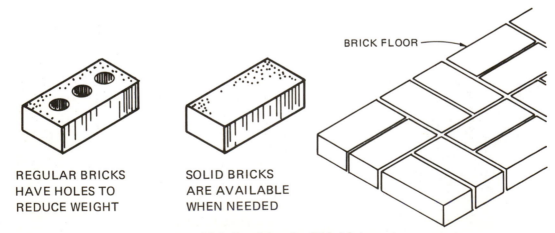

REGULAR BRICKS
HAVE HOLES TO
REDUCE WEIGHT

SOLID BRICKS
ARE AVAILABLE
WHEN NEEDED

BRICK FLOOR

Fig. 23-1 Regular and solid brick types

of walls, chimneys, floors, etc. are made of face brick.

CONCRETE BLOCK

Concrete blocks are rectangular hollow or solid masonry units made of portland cement, selected aggregates, water, and sometimes additives. They are not burned but are steam cured. This process greatly accelerates the hydration (combining with water) and hardening of the cement. The moisture content of the block is controlled so that the unit is ready to be used when cooled. Most concrete blocks are hollow units with two or three cores per unit, figure 23-3. A *hollow block* is defined by the American Society

for Testing Materials (ASTM) as one having a core area of at least 25 per cent of its gross area.

Concrete block is manufactured in modular sizes almost exclusively. Modern vibration block-making machines were developed around 1938, about the time modular coordination studies were being finalized. Concrete block construction is more economical than brick construction. Because of the larger size of the concrete block, buildings can be built more quickly. Many special shapes and face patterns of concrete block are available. Figure 23-4, page 222, gives the sizes of standard units.

Two types of block are made by using different aggregates in the mix. Aggregates of

UNIT	NOMINAL DIMENSIONS			JOINT	MODULAR
DESIGNATION	Thickness	Height	Length	THICKNESS, IN.	COURSING, IN.
STANDARD	4	2 2/3	8	3/8, 1/2	3C = 8
ROMAN	4	2	12	3/8, 1/2	2C = 4
NORMAN	4	2 2/3	12	3/8, 1/2	3C = 8
DOUBLE	4	5 1/3	8	3/8, 1/2	3C = 16
TRIPLE	4	5 1/3	12	3/8, 1/2	3C = 16
SCR	6	2 2/3	12	3/8, 1/2	3C = 8
ECONOMY 12	4	4	12	3/8, 1/2	1C = 4

Fig. 23-2 Table of modular brick units

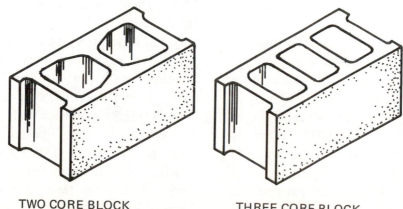

TWO CORE BLOCK   THREE CORE BLOCK

Fig. 23-3 Concrete block core patterns

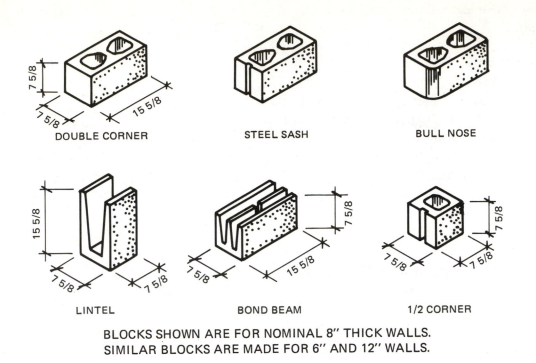

BLOCKS SHOWN ARE FOR NOMINAL 8" THICK WALLS.
SIMILAR BLOCKS ARE MADE FOR 6" AND 12" WALLS.

Fig. 23-4 Standard concrete block shapes

sand and crushed rock are used in *standard concrete blocks. Lightweight concrete blocks* are made using expanded shale, slate, and slag. Lightweight concrete blocks are somewhat more expensive but are preferred for inside masonry work. They have better insulating properties, have a nice texture for exposed work, and are easier to lay. Some lightweight blocks have the same strength as standard units and can be used in load bearing walls.

BONDING

In the early days of masonry construction, most walls were load-bearing. For sizable structures this meant thick walls. Since the masonry units were small, systems of tying the layers together to make them strong were developed, figure 23-5. This technique is called *bonding*. Today the framed structure is used for the most part, and masonry is more of an enclosing wall than a supporting wall. These enclosing walls of structures are called *curtain walls*. These walls are much

thinner and often are made up of a brick facing with tile or block backing, figure 23-6. The traditional bonding methods are largely unnecessary especially with metal joint reinforcement available.

The bonding patterns, however, sometimes are used for decorative purposes even when not structurally needed. Three traditional patterns are the common or American bond, English bond, and Flemish bond, figure 23-7. These bonds were developed for walls eight inches thick or thicker. They are no longer in widespread use because of the recent popularity of veneer masonry. However, Flemish bond can be used for veneering (four inches thick) by using half brick for the headers (bricks with their ends showing.)

Other bonding patterns have been developed in modern times and are used widely. The majority of modern masonry work is laid in the running or stretcher bond, figure 23-8, page 224. The running bond is the simplest and easiest to construct. Since

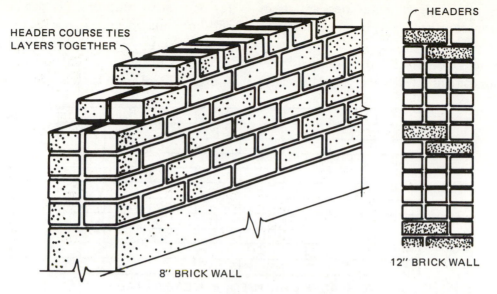

HEADER COURSE TIES LAYERS TOGETHER

HEADERS

8" BRICK WALL

12" BRICK WALL

Fig. 23-5 Bonding masonry walls

3 COURSES OF BRICK LAY UP TO ONE OF BLOCK. SOMETIMES A SPACE IS LEFT BETWEEN THE TWO TIERS WHICH CAN BE LATER FILLED WITH SPECIAL INSULATION.

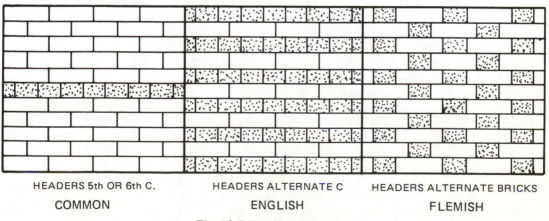

Fig. 23-6 Brick facing with block backup

HEADERS 5th OR 6th C.	HEADERS ALTERNATE C	HEADERS ALTERNATE BRICKS
COMMON	ENGLISH	FLEMISH

Fig. 23-7 Bonding patterns

there are no headers, metal ties are used to tie this masonry to backing or framework, figure 23-9.

Another modern bond pattern is the stacked bond, figure 23-10. It is used basically for decorative purposes. It produces a neat and coordinated appearance in exposed concrete masonry walls and partitions. Many architects prefer it for its crisp pattern and simple control joint details. Since there is no

NO HEADERS ARE USED. INSTEAD METAL TIES OR JOINT REINFORCEMENTS ARE USED.

Fig. 23-8 Running or stretcher bond

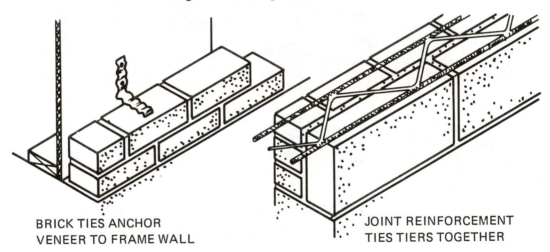

BRICK TIES ANCHOR VENEER TO FRAME WALL

JOINT REINFORCEMENT TIES TIERS TOGETHER

Fig. 23-9 Metal masonry ties

SINCE THE UNITS DO NOT INTERLOCK METAL JOINT REINFORCEMENT IS REQUIRED.

Fig. 23-10 Stacked bond

overlapping of the units in this pattern, steel joint reinforcement is required, figure 23-11. Very careful layout and masonry work is required for stacked bond walls.

LAYOUT

Since the carpenter is responsible for laying out the walls and partitions, the plans must be read with great accuracy. As was pointed out in Unit 6, the nominal thickness of a masonry wall is used, especially if modular drafting standards are followed. These nominal dimensions are 3/8 to 1/2 of an inch greater than the masonry actually is. This means that lines snapped on a floor representing walls and partitions are not the face of the wall but are 3/16 of an inch from

it, figure 23-12. Although more work, the mason can more easily understand double lines representing the nominal thickness of the wall.

The main problem in partition layout comes in setting wall lines for halls, closets, and other small dimensions. These walls usually join some fixed dimension element such as a metal door frame or concrete column, figure 23-13, page 226. Recesses and returns interrupt the pattern of the masonry and must be located to fit the units if possible. The layout must be understood as well. If the mason lays to the line on a short nominal layout, thick head joints will occur, figure 23-14, page 226.

It is a good practice to select a continuous straight line through the building to be

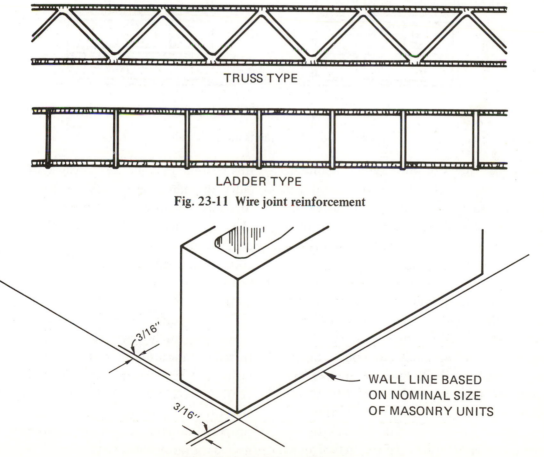

TRUSS TYPE

LADDER TYPE

Fig. 23-11 Wire joint reinforcement

3/16"

3/16"

WALL LINE BASED
ON NOMINAL SIZE
OF MASONRY UNITS

Fig. 23-12 Modular wall layout

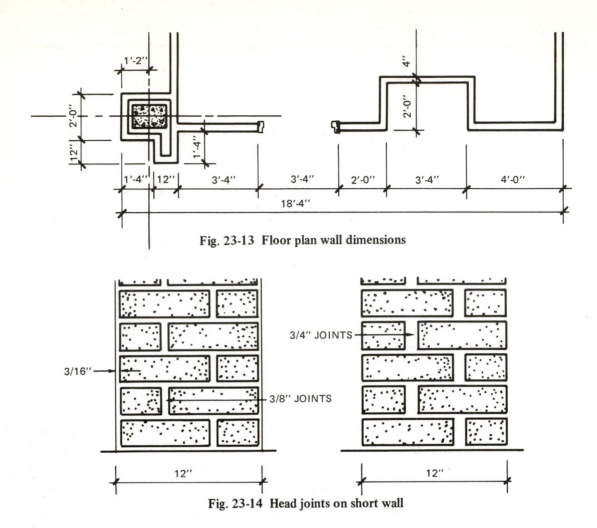

Fig. 23-13 Floor plan wall dimensions

Fig. 23-14 Head joints on short wall

used as a base line for the layout. Column lines from the structural grid are good for calculations but are not continuous after the columns are poured. A center line through the building, when not a column line, is a convenient base line, figure 23-15.

A few architects feel that the actual size of the masonry walls should be dimensioned on the plans. They argue that this eliminates the possible errors that can accumulate with the nominal method. Single wythe (a single vertical section of masonry one unit in thickness) block partitions are 3 5/8 inches or 7 5/8 inches instead of 4 or 8 inches, figure 23-16. This is obvious to the blueprint reader who knows the actual sizes of common masonry units. The accuracy of

this system may be lost because of the difficulty of managing strings of dimension with so many fractions. Compare the two examples in figure 23-16. Exact dimensions at critical points can be shown on details to good advantage, figure 23-17. This allows the main layout to follow the simpler nominal method.

The other extreme is found on some drawings which have very few dimensions. Instead, a modular grid is used to locate most walls and partitions. Usually a wall either centers on a grid line or falls to one side or the other, figure 23-18, page 228. When properly understood, these drawings give the necessary information to build by. A great deal of drafting time is saved in their preparation.

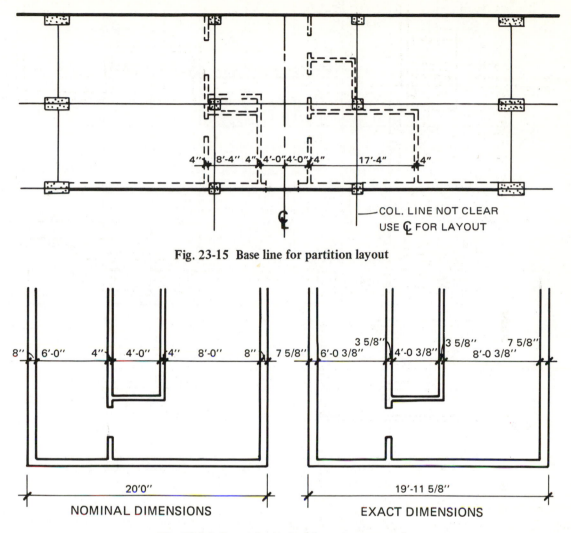

4″ 8′-4″ 4″ 4′-0″ 4′-0″ 4″ 17′-4″ 4″

COL. LINE NOT CLEAR
USE ℄ FOR LAYOUT

℄

Fig. 23-15 Base line for partition layout

8″ 6′-0″ 4″ 4′-0″ 4″ 8′-0″ 8″ 7 5/8″ 6′-0 3/8″ 3 5/8″ 4′-0 3/8″ 3 5/8″ 8′-0 3/8″ 7 5/8″

20′0″

NOMINAL DIMENSIONS

19′-11 5/8″

EXACT DIMENSIONS

Fig. 23-16 Exact & nominal layouts compared

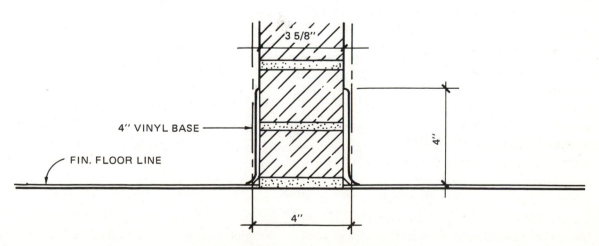

3 5/8″

4″ VINYL BASE

FIN. FLOOR LINE

4″

4″

Fig. 23-17 Masonry exact size on detail

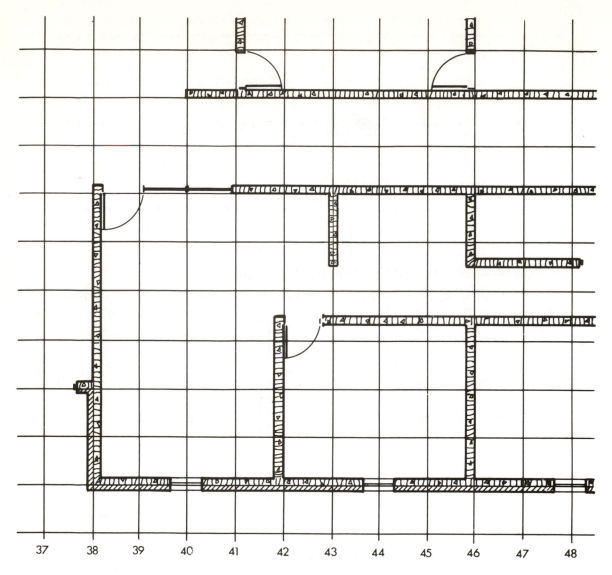

GRID NOTE:
THIS BUILDING IS LAID OUT ON A 4'-0" X 4'-0" GRID. ALL
WALLS AND PARTITIONS THAT ARE NOT DIMENSIONED ARE
CENTERED ON A GRID, HAVE ONE FACE ON A GRID OR HAVE
ONE FACE OR CENTERLINE MIDWAY BETWEEN.

Fig. 23-18 Grid layout with few dimensions

SUMMARY

- Bricks are small solid masonry units made from clay and burned in a kiln.

- Two types of bricks are common brick, used for backup and concealed work, and face brick, used for exposed and decorative work.

- Concrete blocks are large hollow or solid masonry units made of portland cement with suitable aggregate and cured without burning.

- Two types of concrete blocks are made — standard, made with stone aggregate, and lightweight, made with expanded shale, slate, or slag aggregate.

- Thick heavy masonry walls are required for tall buildings when the walls are load bearing.

- The technique of tying masonry units together in thick walls is called bonding.

- Buildings constructed with structural frames are enclosed with thin masonry non-load-bearing walls called curtain walls.

- Bonding patterns sometimes are used for decorative rather than structural purposes.

- The majority of modern masonry work is laid in the running bond with metal ties and joint reinforcement.

- Careful layout is required for masonry work in order to control joint patterns and sizes.

- It is a good practice to establish a continuous base line through the building when laying out interior masonry work.

- Some floor plans locate walls by means of grid lines with very few dimensions.

REVIEW QUESTIONS

1. Give two reasons for the holes usually found in bricks.

2. Sketch a standard modular brick. Dimension the thickness, height, and length according to its nominal size.

3. How many courses of standard modular brick lay up to equal 4'-0" in height?
   a. 24   c. 9
   b. 18   d. 12

4. How many courses of Roman brick lay up to equal 4'-0" in height?
   a. 18
   c. 24
   b. 20
   d. 30

5. What type brick is made to meet standards for color, texture, and strength?

6. Concrete blocks are classed as hollow units when the core area is at least _____ % of the gross area.
   a. 10
   c. 75
   b. 50
   d. 25

7. Based upon the kind of aggregate used in their manufacture, what are the two types of concrete block called?

8. What is the system called for interlocking and tying masonry units together?

9. Sketch brick walls showing the bonding patterns listed below. Shade the headers.
   a. common bond

   b. Flemish bond

   c. stacked bond

10. How far should masonry be held back from nominal layout lines if exact bonding is required?
    a. 3/16 in.   c. 1/2 in.
    b. 3/8 in.   d. 0 in.

11. In what way is a plan that is dimensioned using the exact sizes of masonry units more difficult to work with?

12. How can building plans be drawn with very few dimensions but good results?

Unit 24
Openings and Special Construction

OBJECTIVES

After studying this unit, the student should be able to:

- tell how openings in masonry veneer on wood framing are detailed.
- explain how concrete block lintels and bond beams are constructed.
- describe reinforced masonry walls as used in load-bearing wall construction.

OPENINGS IN WOOD FRAMING

Window and door openings in wood framed buildings are located by their center lines. Masonry in these buildings is used as a finish or veneer. Most of the time it is made of four-inch nominal thickness units such as brick or stone. For economy, a three-inch brick can be used. The masonry veneer is laid with an airspace separating it from the wood for a moisture barrier and for some insulating effect. The airspace is about one inch, figure 24-1.

Window and door frames are set with suitable brick moulding or metal fins to join with the masonry, figure 24-2. The airspace is closed by the window trim when the masonry is laid. Usually a small gap is left between the masonry and the frame for caulking. Steel angles are used as *lintels* (horizontal structural member, used to support the wall above door and window openings). When the cornice detail is brought low enough over the windows and doors, no masonry or steel lintel is required and the head closure is made with wood trim, figure 24-3.

The masonry bonding is not very well coordinated in these buildings. One problem is the one-inch airspace which moves the veneer out from the wood frame. The wood frame usually is modular so the veneer is now off by one inch at each corner. Even with

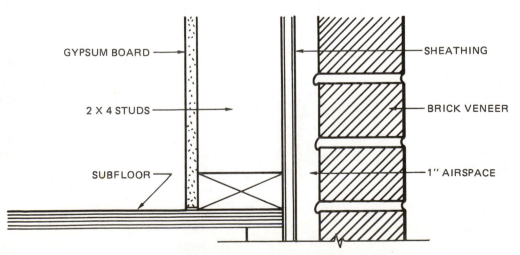

GYPSUM BOARD

2 X 4 STUDS

SUBFLOOR

SHEATHING

BRICK VENEER

1″ AIRSPACE

Fig. 24-1 Brick veneer with airspace

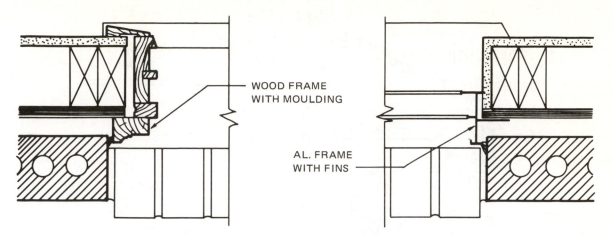

Fig. 24-2 Airspace concealed at frame — plan view

WOOD FRAME
WITH MOULDING

AL. FRAME
WITH FINS

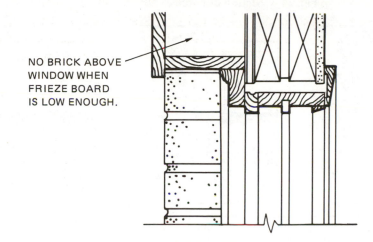

NO BRICK ABOVE
WINDOW WHEN
FRIEZE BOARD
IS LOW ENOUGH.

Fig. 24-3 Head closure with low cornice — elevation view

modular windows and door frames, the dimensions for the masonry have flaws. Fortunately, the mason is able to stretch joints or cut units to make these odd dimensions work.

One detail of importance in the carpenter's control is the frieze board setting. At the top of the wall, the veneer ends behind a frieze board, part of the cornice trim. The carpenter usually sets the frieze board before the veneer is laid. The thickness of the masonry plus the specified airspace must be determined accurately. A section view of the cornice should give this information, figure 24-4, page 234. A change from four-inch to three-inch brick changes this frieze setting

critically. If not moved in for the three inch brick, the brick does not close the airspace at the windows and doors, figure 24-5, page 234. Checking must be done to be sure the brick specified is the one the detail drawings are based upon. If not, then the detail must be revised because the specifications take precedence.

Openings through the floors and roofs for fireplaces and chimneys are also constructed before the masonry is laid. A two inch clearance between the chimney and wood framing is required for safety. The fireplace details give the size of the masonry at various levels. The vertical alignment must be carefully studied since many chimneys have

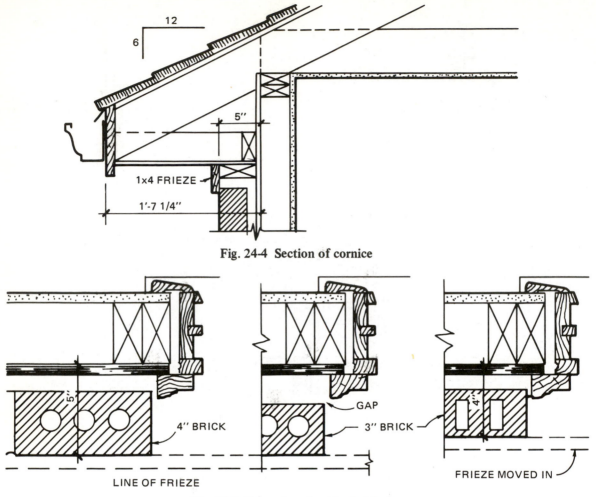

Fig. 24-4 Section of cornice

Fig. 24-5 Frieze board set by brick size

offsets, figure 24-6. Checking the floor plans for location and accuracy before framing the floors is a must.

OPENINGS IN MASONRY WALLS

Window and door openings in masonry walls are dimensioned to their edges rather than to their center lines, figure 24-7. In this way the bonding of the units can be controlled. The openings are called *masonry openings.* Window openings often are built using special jamb units that allow the windows to be installed later, figure 24-8. This is not apparent from the window details since they show the windows fully installed.

Mullions, the trim dividing multiple windows, provide some installation leeway. The finish sills are also laid after the windows are set and help complete the closure. Delaying the setting of the windows until after the walls and roof are complete prevents construction damage to the windows. When windows are set early, glazing is not done until later.

Door openings usually are formed with metal door frames erected before the masonry is laid. After the walls are marked out on the floor slab, the door openings are located. The frames are anchored to the floor and braced securely, figure 24-9, page 236. The

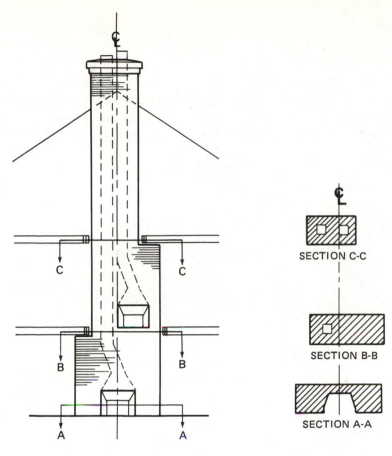

SECTION C-C

SECTION B-B

SECTION A-A

Fig. 24-6 Fireplace chimney with offsets

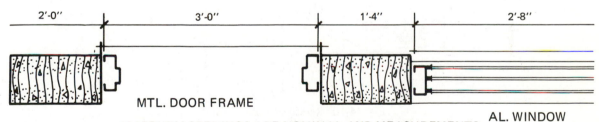

2'-0''   3'-0''   1'-4''   2'-8''

MTL. DOOR FRAME

AL. WINDOW

MASONRY OPENINGS ARE NOMINAL AND MEASUREMENTS ARE TAKEN TO THE CENTER LINE OF THE JOINTS.

Fig. 24-7 Openings in masonry walls

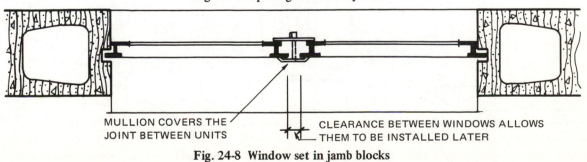

MULLION COVERS THE JOINT BETWEEN UNITS

CLEARANCE BETWEEN WINDOWS ALLOWS THEM TO BE INSTALLED LATER

Fig. 24-8 Window set in jamb blocks

mason now begins laying the walls against the door frames.

Lintels in masonry walls may be made of steel or reinforced concrete. Block walls usually have special lintel blocks which are supported temporarily on a wooden form. The blocks are shaped to provide a space for reinforcing steel and concrete, thus forming a reinforced concrete lintel, figure 24-10.

The elevation of the finish floor line is given in the section details. Sometimes toppings of tile, terrazzo, or some other special material are required. The door frame must be raised to match, figure 24-11. One method

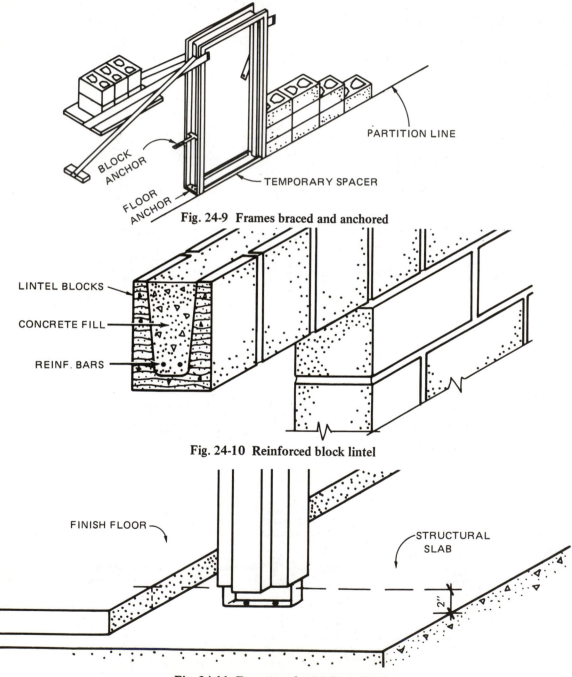

Fig. 24-9 Frames braced and anchored

Fig. 24-10 Reinforced block lintel

Fig. 24-11 Frame set for 2″ floor finish

of controlling the layout in the vertical direction is to mark each column with a grade mark, figure 24-12. This is done with a transit soon after the rough slab is poured. This gives each trade a reference plane to measure from. A rough slab may vary slightly from point to point so it is not dependable for finish work.

SPECIAL MASONRY CONSTRUCTION

Arched openings often are found in modern buildings. They require special forms called *centering*, figure 24-13. This detail usually repeats many times and the centering is moved from arch to arch. The true arch is an ancient form that depends upon compressive forces between many small masonry units for stability. Modern reinforced masonry is superior in strength and resistance to bending forces. Considerably more freedom of design is thus available to architects today and sometimes the arch is purely a decorative form.

Reinforced masonry in load-bearing wall construction is being developed as a competitive system to framed structures. Multistory apartments are one example. The wall is made up of two sections of brick masonry with a space between. The space is for placement of reinforcing steel and concrete, figure 24-14. Floor slabs are typical solid slabs bearing on these walls. The floor plans of

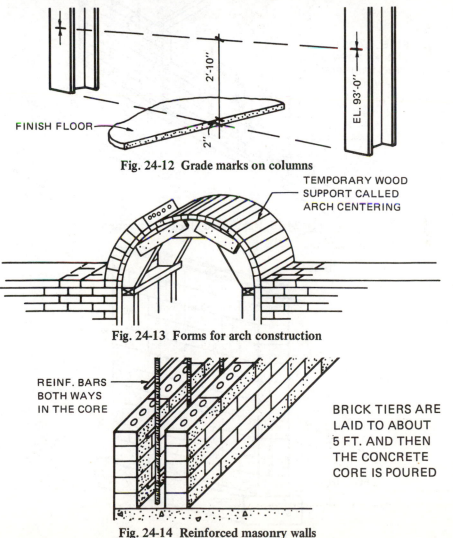

FINISH FLOOR

2'-10"

2"

EL. 93'-0"

Fig. 24-12 Grade marks on columns

TEMPORARY WOOD
SUPPORT CALLED
ARCH CENTERING

Fig. 24-13 Forms for arch construction

REINF. BARS
BOTH WAYS
IN THE CORE

BRICK TIERS ARE
LAID TO ABOUT
5 FT. AND THEN
THE CONCRETE
CORE IS POURED

Fig. 24-14 Reinforced masonry walls

these buildings would be easily recognized because of the lack of columns of any kind.

Reinforcement of concrete block buildings to help them resist earthquake and wind stresses is required in certain areas of the country. Two special techniques are used in these buildings. One consists of locating vertical rods in the cells of the block at frequent and special places. The cells containing the steel bars are filled with concrete forming reinforced concrete studs. Dowels are set in the footing at carefully located points to allow the studs to be anchored.

Designs vary but studs usually are located at each corner, beside the openings, and at 24 inch intervals along the walls, figure 24-15.

The other special technique occurs at the top of the wall and at floor lines for buildings with more than one story. It is called a *bond beam* and is constructed using special U-shaped blocks for the entire course. Steel bars and concrete are placed in the hollow space forming a continuous reinforced beam. The stud reinforcement extends into the bond beam effectively tying the entire wall to the footing, figure 24-16.

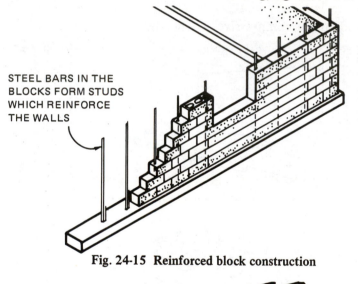

STEEL BARS IN THE BLOCKS FORM STUDS WHICH REINFORCE THE WALLS

Fig. 24-15 Reinforced block construction

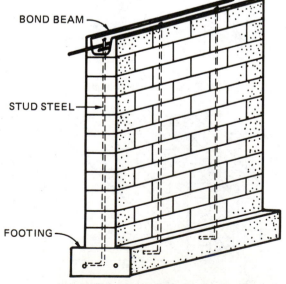

BOND BEAM

STUD STEEL

FOOTING

Fig. 24-16 Bond beam tied to footing

SUMMARY

- Window and door openings in wood framed buildings are located by their center lines.

- Masonry veneer is usually made of nominal four inch thick material spaced one inch out from the wood construction.

- Masonry bonding is difficult in wood veneered buildings because of the one inch extra distance caused by the airspace.

- The frieze board is usually set by the carpenter before the veneer is laid and must be located carefully.

- A two inch clearance around masonry chimneys is required when framing openings through wood floors and roofs.

- Masonry openings are dimensioned to their edges in masonry construction.

- Metal door frames are located, anchored to the floor, and braced securely before the masonry is laid.

- Lintels above openings in masonry walls may be made of structural steel or reinforced concrete.

- Arched openings in masonry walls are built on special forms called centering.

- Modern reinforced masonry can be used for load-bearing structures and compete favorably with framed structures.

- Concrete block buildings are reinforced with steel bars set vertically like studs and horizontally in bond beam courses.

REVIEW QUESTIONS

1. The airspace between masonry veneer and wood framed walls should be about _____ inches wide.

   a. 1/2   c. 2

   b. 1   d. 4

2. When wood windows are used, the airspace is covered by

   a. wood brick mouldings.   c. metal flashing.

   b. caulking.   d. the window frame.

3. How is the brick veneer supported when it is laid above a window or door opening?

4. How is the location of the frieze board determined when set before the veneer is laid?

5. To be safe, a clearance of _____ inches should be provided between wood framing and chimneys.
   a. 1/2   c. 2
   b. 1   d. 3

6. Why are openings in masonry walls dimensioned to their edges rather than to their center lines?

7. Are brick window sills laid before or after the windows are installed?

8. When are steel door frames usually set in masonry construction?

9. Name two types of lintels used in masonry construction.

10. Sketch a section of a reinforced concrete block lintel.

11. What are the forms used to support masonry arches during their construction called?

12. How can steel bars be used with brick masonry to make multistory column-free buildings?

13. Describe two techniques used to reinforce concrete block buildings.

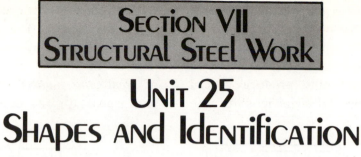

Unit 25
Shapes and Identification

OBJECTIVES

After studying this unit, the student should be able to:

- tell what work is properly classified as structural steel work.

- explain why a new system of shape designations was adopted in the AISC Manual.

- give the meaning of beam designations.

CLASSES OF STRUCTURAL STEEL

Structural steel work, as described in the C. S. I. Section 05100, consists of columns, beams, girders, bases, and minor parts. Steel work not classed as structural steel is concrete rebars, metal decking, and other metal parts. Structural steel work is a specialized phase of construction. For such construction (according to some specs,) only qualified firms with at least five years of experience are to be used.

Steel is now the most versatile of all structural materials. It is very strong, workable, weldable, and easy to buy. Its properties are uniform and known in advance. Construction steels are made in a number of classes, with different properties. To provide uniform properties within each class, regardless of the mill that makes it, specifications prepared by ASTM are used. ASTM is the American Society for Testing and Materials, and reference to its specifications is found in almost every construction project.

Besides different strength properties, some classes of structural steel are corrosion resistant and need not be painted. This class sometimes is called *weathering steel*.

The workhorse of the construction steels is now A36 and it is also the lowest in price. Until the late 1950's, A7 steel was the most frequently specified type but it is now replaced by A36. Several other classes of steel are available which have greater strength than A36. They are known as *high-strength steels*. They are particularly useful where the weight of the structure is the principal load carried. Very tall buildings and long bridges are examples of structures with large dead loads (weight of structure).

ROLLED SHAPES CLASSIFIED

Construction steels are classed as either plates, bars, or shapes. *Plates* are large flat pieces of steel of various thicknesses. They are cut into parts for fabrication of plate girders and column bases to name two examples, figure 25-1, page 242. *Bars* are smaller pieces of flat steel eight inches or less in width. They are available in a number of sizes and are used widely. They include plain rods of either round, square, or other cross section, figure 25-2, page 242. *Shapes* may be angles, channels, S-beams, Wide-Flange sections, Tee sections, or any section that can be rolled.

These rolled shapes are formed by passing a hot billet or bar of steel through special rolls which gradually squeeze it into the shape desired, figure 25-3.

Although steel mills produce special shapes of their own design, most of the shapes rolled conform to standardized forms and dimensions. The dimensions of these standard shapes appear in the American Institute of Steel Construction (AISC) *Manual of Steel Construction.* The Manual also contains specs for the design, fabrication, and

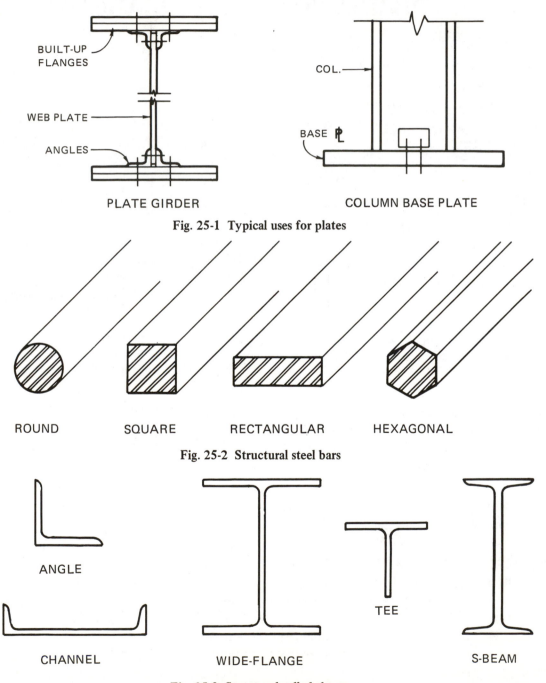

Fig. 25-1 Typical uses for plates

Fig. 25-2 Structural steel bars

Fig. 25-3 Structural rolled shapes

erection of structural steel for buildings. These specs are considered *the* standards of the industry and usually are included by reference in building plans.

The new seventh edition of the AISC Manual contains a new system of designations of structural shapes. The chief reason for this change is to make the designations easier to work with in computer applications. Because of long standing usage, the older designations may still appear on drawings. The following descriptions and comparison of old and new designations should be mastered if the blueprint reader expects to be able to interpret structural drawings correctly.

ANGLES

Angles, now called the L-shape, may have equal legs or unequal legs, figure 25-4.

Both legs are always of the same thickness. An example of the new designation for angles is L 6 x 4 x 5/8. This signifies an unequal leg angle, one leg being 6 inches, the other leg 4 inches. The thickness of both legs is 5/8 of an inch. The former designation was the same except for the first character, ∠ 6 x 4 x 5/8. A special symbol was used but now an upper case L has been substituted. Drawings may not show much difference, but any printed material uses the new standard capital L.

CHANNELS

Channels, now called C-shapes, are sections with a pronounced slope on their inner flange surfaces, figure 25-5. There are two types of channels listed in the Manual. They are American Standard channels and Miscellaneous channels. The Miscellaneous channels

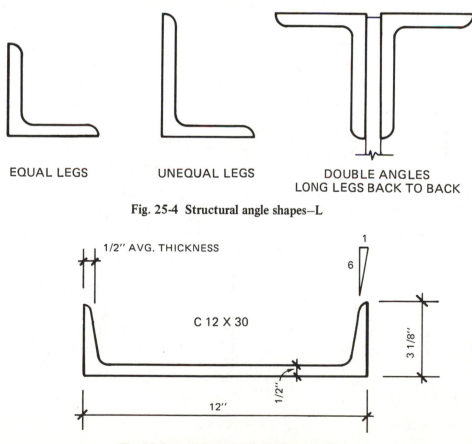

EQUAL LEGS   UNEQUAL LEGS   DOUBLE ANGLES
LONG LEGS BACK TO BACK

Fig. 25-4 Structural angle shapes—L

1/2″ AVG. THICKNESS

C 12 X 30

3 1/8″

1/2″

12″

Fig. 25-5 American standard channel—C

now include the Jr channels of previous editions.

Typical examples of the new designation are C 12 x 20.7 for American Standard channels and MC 12 x 10.6 for Miscellaneous channels. This sample American Standard channel is 12 inches deep and weighs 20.7 pounds per foot. The Miscellaneous channel is also 12 inches deep but weighs 10.6 pounds per foot. The first number following the symbol is always the depth. The second number is the weight per foot. The former designations for these channels were 12 [20.7 and 12 Jr [10.6 respectively. Note that the depth was given first, then a special symbol resembling the channel. The special symbol has been discarded and a standard upper case letter C is now used.

I-SHAPED BEAMS

I-shaped beams may be American Standard, Wide-Flange, or Miscellaneous sections, figure 25-6. The American Standard shape has narrow flanges. The inside surface of the flanges has a rather steep slope like that of the channels. The new symbol for these sections is S. They are called S-shapes. An example is S 24 x 100. This is an I-shaped Standard beam (S beam) with a 24-inch nominal depth and weighing 100 pounds per foot. The old designation was 24 I 100

for the same member. No special symbol was used, but the I was changed to S and placed first in the new system.

The Wide-Flange shapes have wider flanges and thinner webs than the S-shapes. The flanges are of constant thickness. The Wide-Flange section was developed to be a more efficient shape in resisting bending stresses. These Wide-Flange shapes are used much more in building structures than are the S-shapes. The Wide-Flange shape is now simply called a W shape. An example of the new designation is W 24 x 76. This is a 24 inch deep I-shaped beam with wide flanges and weighing 76 pounds per foot. The older designation was 24 WF 76. Here again a special symbol, WF , was used. It is now discarded and W is substituted. In all these new designations a letter symbol comes first, the depth second, and the weight per foot last.

Miscellaneous I-shaped sections are designated by the letter M. These are all the shapes that cannot be classified as W or S. M shapes are rolled by a limited number of mills. Examples of M shapes are M 8 x 18.5 and M 10 x 9. These were formerly called 8 M 18.5 and 10 Jr 9.0.

TEE SECTIONS

Structural Tees are cut from standard W, S, and M beam sections, figure 25-7. The

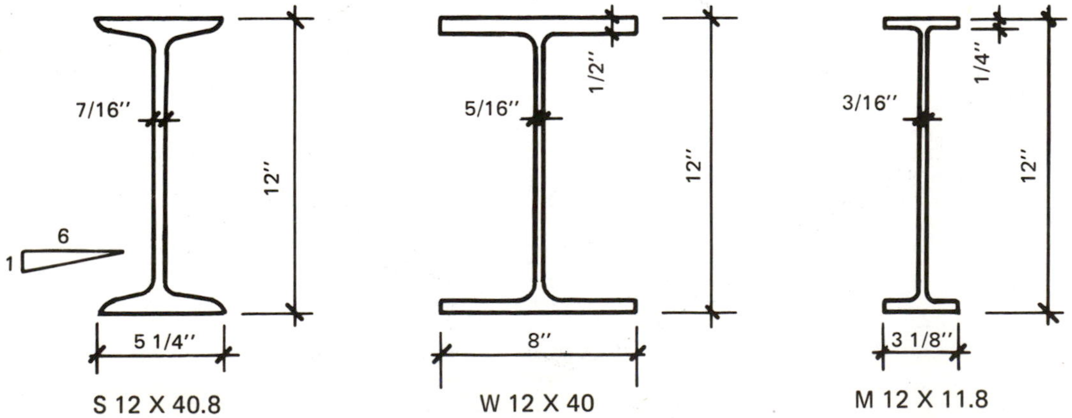

S 12 X 40.8   W 12 X 40   M 12 X 11.8

Fig. 25-6 I-shaped beams—S—W—M

beam is sheared down the center of the web, making two equal Tee sections. The new designation uses the symbol of the beam from which the Tee is cut with an upper case T as the prefix. This is followed by the depth and weight per foot just like the other beams (ST 12 x 50, WT 12 x 38, MT 5 x 4.5). The MT 5 x 4.5 is a Tee section cut from an M 10 x 9 with half the depth and half the weight. The older designations were ST 12 I 50, ST 12 W 38, and ST 5 Jr 4.5 for the examples given.

PLATES

Plates are designated by their cross sectional dimensions, such as PL 1/2 x 18. The only change is in placing the thickness before the width. The older designation was PL 18 x 1/2. A complete note would include the length as well and would be PL 1/2 x 18 x 1'-8", for example. This means that the plate is 1/2 inch thick, 18 inches wide, and one foot eight inches (20 inches) long, figure 25-8.

BARS

There is no change in the bar designations. Typical examples are Bar 1 ⊡ (square bar), Bar 1 1/4 ⊘ (round bar), Bar 2 1/2 x 1/2 (flat bar), figure 25-9, page 246. The symbols ⊡ and ⊘ are used on drawings and simply mean square or round.

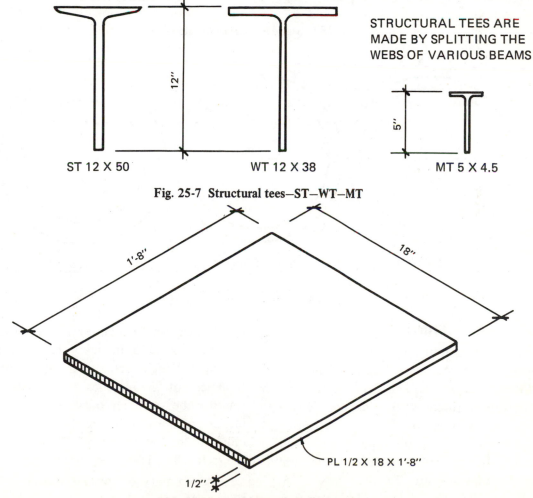

STRUCTURAL TEES ARE MADE BY SPLITTING THE WEBS OF VARIOUS BEAMS

ST 12 X 50   WT 12 X 38   MT 5 X 4.5

Fig. 25-7 Structural tees—ST—WT—MT

PL 1/2 X 18 X 1'-8"

Fig. 25-8 Standard plate designation

PIPE

Although not formed like the hot rolled sections just described, pipe is an important shape in structural steel. Pipe is a round hollow section that is very efficient when used as a column, figure 25-10. Pipe for structural work is made in three strengths and many sizes. The size is the nominal inside diameter of the pipe plus the strength class. The wall thickness is increased for more strength. No change has been made in the pipe size designation. Examples of the three pipe strengths are Pipe 4 Std, Pipe 4 x-strong, Pipe 4 xx-strong. These are read 4 inch standard weight pipe, 4 inch extra-strong pipe, and 4 inch double-extra strong pipe respectively.

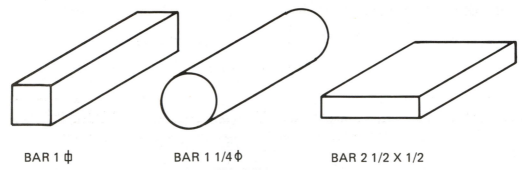

BAR 1 ⏀   BAR 1 1/4 ⏀   BAR 2 1/2 X 1/2

Fig. 25-9 Standard bar designations

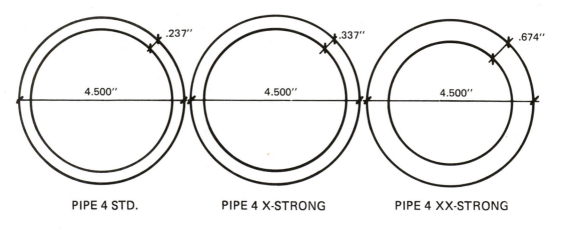

PIPE 4 STD.   PIPE 4 X-STRONG   PIPE 4 XX-STRONG

Fig. 25-10 Pipe sections compared

STRUCTURAL TUBING

Structural tubing is made in square, rectangular, and circular shapes. The outside dimensions are used to describe the section together with the wall thickness. Structural tubing is a good choice for exposed structures with the square types being easier to detail, figure 25-11. Some structural tubing designations are TS 4 x 4 x .375, TS 5 x 3 x .375, TS 3 OD x .250. The first is a 4 inch square structural tube with a wall thickness of .375 inches. The second is a 5 inch x 3 inch rectangular tube with a wall thickness of .375 inches. The last one is a round tube with an outside diameter of 3 inches and a wall thickness of .25 inches. The older method used the word tube instead of the letters TS (Tube 4 x 4 x .375).

A summary of the old and new designations as described above is given in figure 25-12.

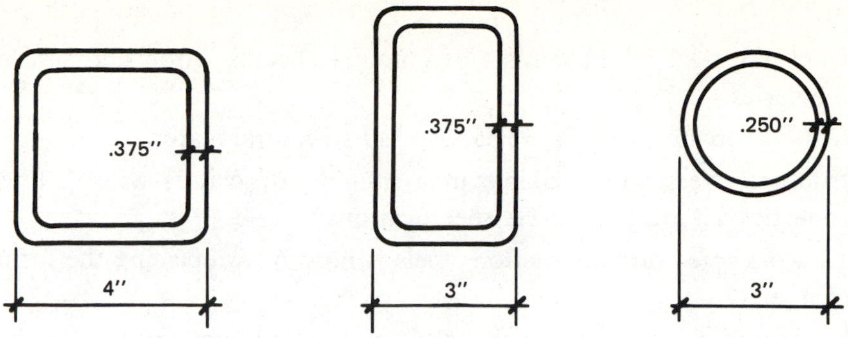

TS 4 X 4 X .375 TS 5 X 3 X .375 TS 3 OD X .250

Fig. 25-11 Structural tubing

New Designation	Type of Shape	Old Designation
W 24 x 76 W 14 x 26	W shape	24 WF 76 14 B 26
S 24 x 100	S shape	24 I 100
M 8 x 18.5 M 10 x 9 M 8 x 34.3	M shape	8 M 18.5 10 JR 9.0 8 x 8 M 34.3
C 12 x 20.7	American Standard Channel	12 [20.7
MC 12 x 45 MC 12 x 10.6	Miscellaneous Channel	12 x 4 [45.0 12 JR [10.6
HP 14 x 73	HP shape	14 BP 73
L 6 x 6 x 3/4 L 6 x 4 x 5/8	Equal Leg Angle Unequal Leg Angle	∠6 x 6 x 3/4 ∠6 x 4 x 5/8
WT 12 x 38 WT 7 x 13	Structural Tee cut from W shape	ST 12 WF 38 ST 7 B 13
ST 12 x 50	Structural Tee cut from S shape	ST 12 I 50
MT 4 x 9.25 MT 5 x 4.5 MT 4 x 17.15	Structural Tee cut from M shape	ST 4 M 9.25 ST 5 JR 4.5 ST 4 M 17.15
PL 1/2 x 18	Plate	PL 18 x 1/2
Bar 1ф Bar 1 1/4Φ Bar 2 1/2 x 1/2	Square Bar Round Bar Flat Bar	Bar 1 ф Bar 1 1/4 Φ Bar 2 1/2 x 1/2
Pipe 4 Std. Pipe 4 X-Strong Pipe 4 XX-Strong	Pipe	Pipe 4 Std. Pipe 4 X-Strong Pipe 4 XX-Strong
TS 4 x 4 x .375 TS 5 x 3 x .375 TS 3 OD x .250	Structural Tubing: Square Structural Tubing: Rectangular Structural Tubing: Circular	Tube 4 x 4 x .375 Tube 5 x 3 x .375 Tube 3 OD x .250

Fig. 25-12 Summary of old and new hot-rolled structural steel designations

SUMMARY

- Structural steel work consists of columns, beams, girders, bases, and accessories.
- Steel is probably the most versatile of all structural materials.
- Structural steels are available in a number of classes with different properties, all meeting ASTM specifications.
- The workhorse of construction steels is now A36 replacing the former A7 steel.
- High strength steels are available for very tall buildings and long bridges.
- Construction steels are classed as either plates, bars, or shapes.
- Shapes may be angles, channels, S-beams, Wide-Flange sections, Tee sections, or other hot rolled sections.
- The American Institute of Steel Construction publishes the *Manual of Steel Construction* which gives data on standardized shapes and specifications for the design, fabrication, and erection of structural steel.
- The AISC manual gives a new system of designating shapes in order to make these notations easier to work with in computer applications.
- Two important notation changes are the symbols used and the order of numerical data.
- American Standard I-beams are now called S-sections as in S 24 x 100.
- Wide-Flange beams are now called W-sections as in W 24 x 76.
- Miscellaneous I-shaped sections formerly called M and Jr beams are now called M-sections as in M 10 x 9.
- Structural Tees are now designated using the symbol for the section from which they are cut followed by the letter T as in WT 12 x 38.
- Plates are designated by their cross sectional dimensions as in PL 1/2 x 18.
- Bar designations use special symbols ▢ (square) and ◯ (round) with the size as in Bar 1 ▢ and Bar 1 1/4 ◯.
- The three strengths of pipe are designated Std., X-strong, and XX-strong as in Pipe 4 Std., Pipe 4 X-strong, and Pipe 4 XX-strong.
- Structural tubing is designated by the outside dimensions and the wall thickness as in TS 5 x 3 x .375.

REVIEW QUESTIONS

1. Which of the following is not classed as structural steel work in CSI, Section 05100?

   a. columns.   c. girders.

   b. beams.   d. concrete rebars.

2. What is the most versatile of all structural materials?

3. What are structural steels that are corrosion resistant and never need painting called?

4. Which ASTM classification of steel is now the most used in construction?
   a. A7.
   c. A440.
   b. A36.
   d. A588.

5. Sketch sections of these shapes which are commonly rolled in structural steel.
   a. angle.

   b. channel.

   c. wide-flange.

6. Why has the system for designating structural steel shapes been recently changed in the AISC Manual?

7. Give the new symbol now being used for each of the following.
   a. angle.

   b. American Standard Beam.

7. c. Wide-Flange Beam.

    d. channel.

8. Compare with sketches the difference between S and W shapes in steel beams.

9. How are structural Tees made?

10. Give the meaning of the following beam designations.
    a. W 24 x 76

    b. S 24 x 100

Unit 26
Structural Steel Drawings

OBJECTIVES

After studying this unit, the student should be able to:

- explain how columns are listed in a column schedule.
- tell how beams are shown on a structural framing plan.
- describe an open web steel joist and give a sample designation.

STRUCTURAL STEEL DRAWINGS

Drawings for steel buildings are different from those used for concrete buildings. They are much more symbolic using simple time-saving conventions. The basic structural drawings furnished with the contract documents sometimes are referred to as the *engineering drawings*. They are prepared by the structural engineer and show the essential features of the structure.

Shop drawings are prepared by the fabricator from these drawings. They are very detailed showing every piece, hole, and cut. They also include erection plans with each piece numbered for easy identification. It is important to be able to read the engineering drawings for related general information. The following material on structural steel drawings is given for that purpose.

FOUNDATION

Foundation plans for steel buildings are similar to those for reinforced concrete buildings except for the steel columns. The foundation plan is laid out by means of a grid (Unit 11) which identifies each column, figure 26-1, page 252. The column base plate details explain the method of joining the column to the foundation, figure 26-2, page 252. Pile foundations and steel grillage foundations were discussed in Unit 14. Figure 14-7 shows steel grillage foundation details. Steel framing is used for very tall buildings as well as simple one-story buildings. Foundation details vary according to the loads and soil conditions encountered.

COLUMNS

Steel columns for multistory buildings are listed in a column schedule, figure 26-3, page 253. Each column is identified by the grid coordinates which locate it on the foundation plan and the floor plans. Similar columns are grouped together and a single heavy solid line is drawn to represent that column. It is a usual practice to make the columns in sections extending two stories in height. The splice is made about three feet above the floor line for convenience. All critical dimensions, column sizes, and locations in the structure are shown on the column schedule. Column splices are detailed for typical cases. From this information, plus the specs and the framing plans for each floor, the steel fabricator prepares the shop drawings.

For one-story or simple buildings, the columns may not be scheduled as above. Instead, they are identified on the framing plan by a note or other designation, figure 26-4, page 253. The identification is referenced to the column by an arrow drawn at 45° to the beam lines. The column location usually is dimensioned directly without a grid.

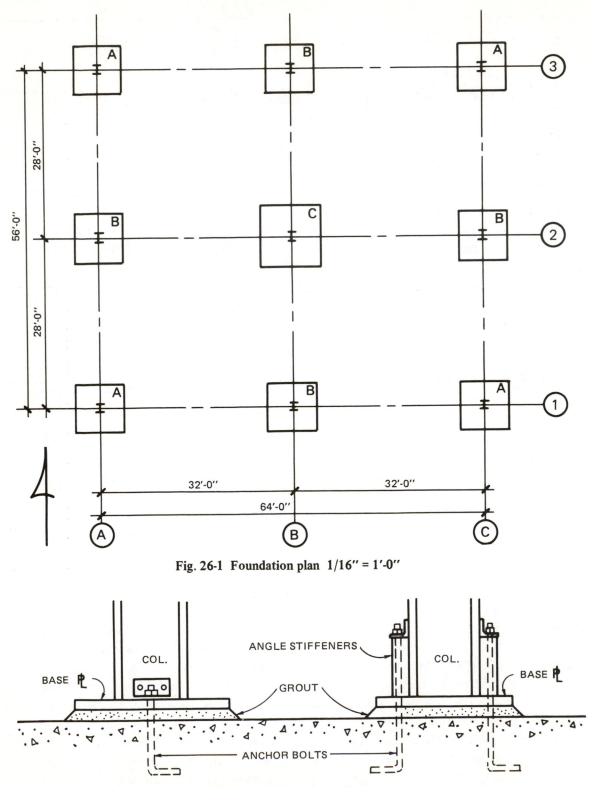

Fig. 26-1 Foundation plan 1/16″ = 1′-0″

Fig. 26-2 Column base plates

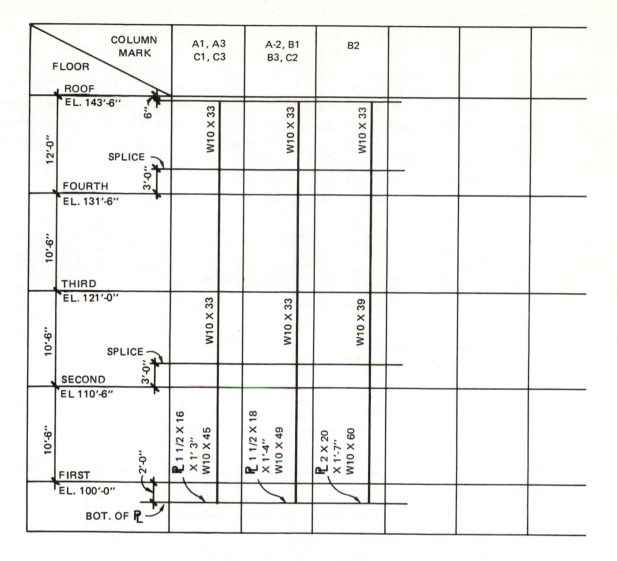

Fig. 26-3 Column schedule

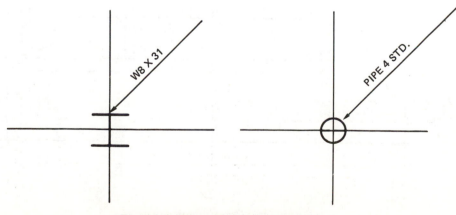

Fig. 26-4 Columns identified by note

FRAMING PLANS

Floor framing plans are drawn for each different framing situation. Most multi-story buildings have several floors framed alike. A note on a typical plan tells which floors are similar. The shop drawings take advantage of this duplication of parts, but for erection purposes each floor level carries an identifying mark.

The framing plans are laid out on the same grid as the foundation plan, and the columns are shown in position. The beams, girders, joists, bridging, and related information are recorded on the framing plans. Main structural members are shown as heavy solid lines, figure 26-5. When a beam or joist frames into a girder and is not continuous over it, its line symbol stops short of the support. The same convention is used when members frame into columns. At first glance this might seem like careless drafting, but it helps identify the extent of each beam. If the lines continue unbroken through the plan, the detailer does not know where to cut the

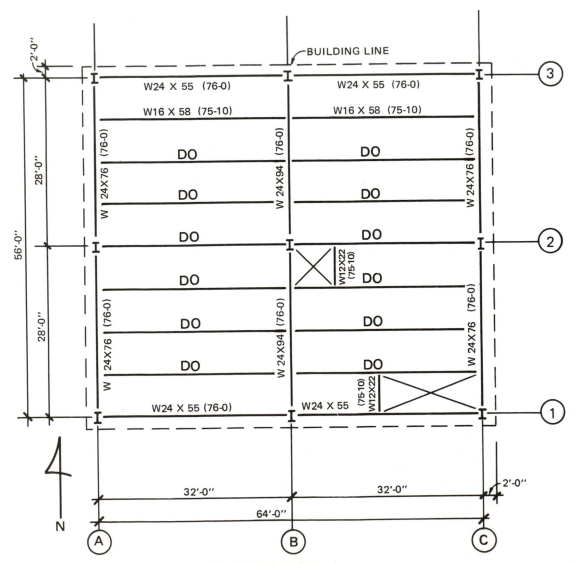

Fig. 26-5 Floor framing plan

members. On the other hand, when the line does pass a member unbroken, a special case is indicated. A common example would be a cantilever beam, figure 26-6.

Each member is identified using the designations as discussed in Unit 25. By each beam a designation is given in parentheses stating the elevation of the top of the member. This may be given as the elevation value or as the distance above or below the floor line, figure 26-7.

A common member not given in Unit 25 is the *open-web steel joist* or *bar joist.* These members are fabricated in standard sizes. They really are trusses made up of angles, round bars, etc., welded together, figure 26-8, page 256. They are available in J-Series

and H-Series sizes depending upon the grade of steel used in their manufacture. J-series indicates steel with a minimum yield of 36,000 p.s.i. H-series indicates steel with a minimum of 50,000 p.s.i. Several groups of designs are manufactured based upon the J and H-Series materials. The range in sizes and spans is shown in figure 26-9, page 256. Typical designations of steel joists are 12 J5 and 26 H8. The first number is the depth of the joist. The letter J or H indicates the type steel used. The last number indicates the chord size, figure 26-8.

When several members are the same size and close enough together to be obvious, the symbol DO is used, figure 26-10, page 256. This means ditto or the same. Every beam or

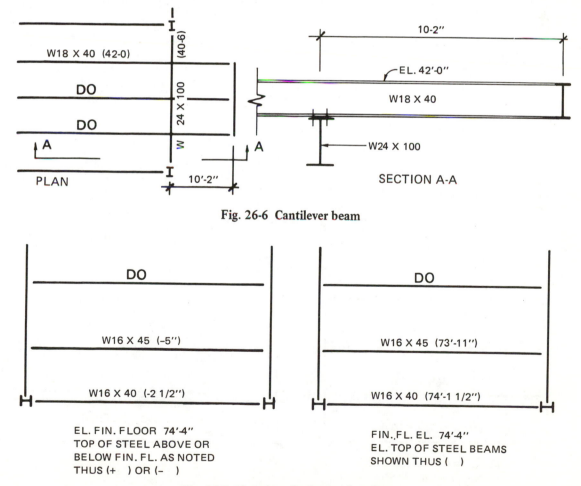

Fig. 26-6 Cantilever beam

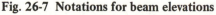

Fig. 26-7 Notations for beam elevations

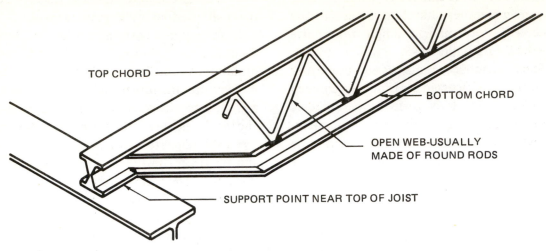

TOP CHORD

BOTTOM CHORD

OPEN WEB-USUALLY
MADE OF ROUND RODS

SUPPORT POINT NEAR TOP OF JOIST

Fig. 26-8 Typical open web steel joist

SERIES	SIZE RANGE	SPAN IN FEET
J	8J3 TO 30J11	8 TO 60
H	8H3 TO 30H11	8 TO 60
LJ	18LJ02 TO 48LJ19	21 TO 96
LH	18LH02 TO 48HL17	21 TO 96
DLJ	52DLJ12 TO 72DLJ20	61 TO 128
DLH	52DLH10 TO 72DLH19	61 TO 128

LJ & LH LONGSPAN JOISTS, DLJ & DLH DEEP LONG SPAN JOISTS

Fig. 26-9 Open web steel joists

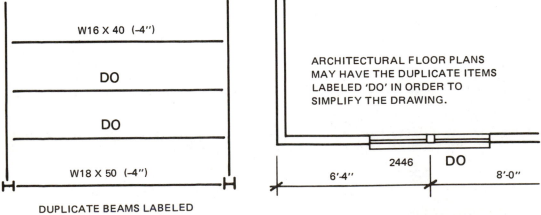

W16 X 40 (–4″)

DO

DO

W18 X 50 (–4″)

DUPLICATE BEAMS LABELED
'DO'. SEE 26-5, 26-7 ALSO

ARCHITECTURAL FLOOR PLANS
MAY HAVE THE DUPLICATE ITEMS
LABELED 'DO' IN ORDER TO
SIMPLIFY THE DRAWING.

2446   DO

6′-4″   8′-0″

Fig. 26-10 Notation of similar members

joist in the series marked DO is the same as the member before it. Architects use this DO symbol on floor plans and details in a similar way. DO is pronounced doe or dough, not do.

Steel joists also are indicated in another way on many drawings. Since these members are spaced closely, two to five feet on center, a group of joists may be indicated by a note. The joists may not be drawn at all except for the first and last ones in a group, figure 26-11. Any oddly spaced joists are drawn also.

Bridging is shown as a dotted line for each run of bridging. Steel joists require bridging much like wood joists. It may be two continuous horizontal steel members, one attached to the top and one to the bottom. Diagonal bridging also is used in certain cases, figure 26-12.

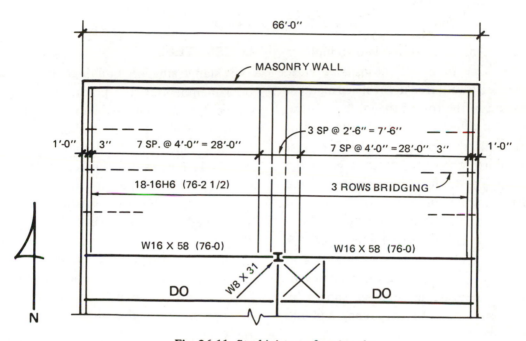

Fig. 26-11 Steel joists on framing plan

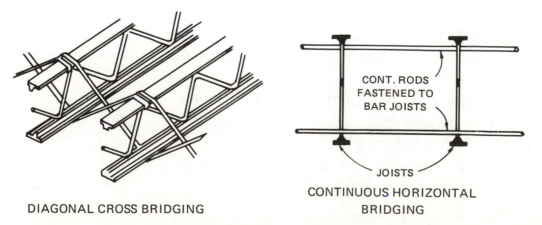

Fig. 26-12 Open web joist bridging

MARK	SPAN	SECTION	FOR
L1	5' - 8"	2 ⌐S 3 1/2 X 3 X 5/16 X 6'-4"	DOORS
L2	7' - 0"	3 ⌐S 6 X 4 X 5/16 X 7'-11"	WINDOW
L3	4' - 0"	3 ⌐S 4 X 3 1/2 X 5/16 X 5'-0"	DO
L4	6' - 0"	1 C 6 X 8.2 X 7'-4" w/ ⌐ 1/4 X 12 X 5'-11 1/2"	DO
L5	3' - 4"	2 ⌐S 3 1/2 X 3 X 1/4 X 4'-0"	DO
L6	3' - 8"	3 ⌐S 3 1/2 X 3 X 1/4 X 4'-4"	DO

Fig. 26-13 Lintel schedule

Two dimension lines are used to indicate steel joists. One indicates the total number and type of joists required in the space marked. The elevation of the top of the joists is indicated also. The other is a string of dimensions beginning with the spacing of the first joist from the starting point. Next is given the number and size of the regular spaces and the total distance this amounts to. Any odd spaces are also indicated. Finally, the distance from the last joist to the end point is shown. This gives excellent layout data as well as a means of checking the accuracy of the plan. The bridging runs are shown by dotted lines but only the beginning and end are drawn, figure 26-11.

LOOSE STEEL

Many smaller buildings use structural steel members at key points even when constructed of masonry or wood. Because these members are not connected in a framework, they are called 'loose.' A good example is the *lintel*. This is a steel member used to support masonry over an opening. If there are several alike, a schedule may be used and the lintels are labeled on the plan, figure 26-13. In other cases the member is identified by a note near the lintel. A single beam or girder to support the floor or roof construction also might be called a loose piece of steel. Section drawings are necessary to give the setting details for these members.

SUMMARY

- Drawings for steel buildings are more symbolic than drawings for concrete buildings.

- Contract drawings for the structure are sometimes called engineering drawings.

- Shop drawings are prepared by the fabricator and are very detailed.

- Foundation plans for steel buildings are similar to those for reinforced concrete buildings.

- Steel columns for multistory buildings are listed in a column schedule.

- Steel columns usually are spliced at two-story intervals.

- For simple buildings, columns may be identified by note at each column.

- Framing plans are laid out on the same grid as the foundation plan.

- Beams are indicated with heavy solid lines which stop short of the members into which they are framed.

- The elevation of the top of steel beams is given in parentheses by each member on framing plans.

- Open-web steel joists are available in J-Series and H-Series sizes depending upon the grade of steel used in their manufacture.

- The symbol DO is used on drawings to mean ditto or the same.

- Lines of bridging are shown as dotted lines on the framing plans.

- Isolated steel beams and lintels are called loose steel members because they do not frame into steel framework.

REVIEW QUESTIONS

1. In steel framed buildings which drawings are the most detailed?

   a. the engineering drawings.   c. the shop drawings.
   b. the framing plans.   d. the structural drawings.

2. Which drawings are more important to the builder?

   a. the engineering drawings.
   b. the shop drawings.

3. How are columns located and identified on plans for steel framed buildings?

4. Steel columns in multistory buildings are usually _____ stories in height.

   a. one.   c. three.
   b. two.   d. four.

5. What does it mean when the line representing a beam on a framing plan stops short of the column or girder that supports it?

6. What information is given in parentheses beside beams on framing plans?

7. Draw a picture of a typical bar joist.

8. How are the parts of a bar joist held together?

9. What does the joist designation 12 J5 mean?

10. What does the symbol DO mean when used on drawings?

11. Why are some steel members called loose steel?

12. Answer the following questions by studying figure 26-11.
    a. How many bar joists are required?

    b. How deep are these bar joists?

    c. Will the bottom of the joists be above or below the bottom of the supporting beam?

    d. What is the weight per foot of the supporting beam?

    e. What section is used for the supporting column?

Unit 27
Fastening Systems

OBJECTIVES

After studying this unit, the student should be able to:

- explain why high-strength bolts are more popular than hot rivets in structural steel erection work.
- tell how high-strength bolts can be identified.
- draw a welding symbol for a typical fillet weld.

FASTENERS FOR STRUCTURAL STEEL MEMBERS

The selection of each steel member is based upon the properties of its section as related to stresses caused by the loads put on it. In order for these members to do their jobs, they must be properly connected. The connections are included in structural steel drawings and specs. They must be understood, first, in order for the fabricator to give a bid on the building. Second, they must later be detailed in the shop drawings so that the steel can be prepared for erection.

RIVETS

Until about 1950, rivets were the most important fasteners used in structural steel work. Almost all shop fabrication as well as some field erection, was done with rivets.

Rivets used in structural steel work are hot driven. The rivet, with a button head formed on one end, is heated and placed in holes through the steel parts, figure 27-1. The other head is formed by a pneumatic hammer called a riveter. Because of the noise of the riveting hammer, rivets have given way to bolts and welding for field erection. Welding is fast becoming the preferred method for shop fabrication as well.

Rivets are a dependable, long-lasting fastener as witnessed by the many older buildings and bridges still in service. When the rivet is driven, the hot metal is forced to fill the hole even when slightly misaligned, figure 27-1c. Because of its temperature, the rivet is

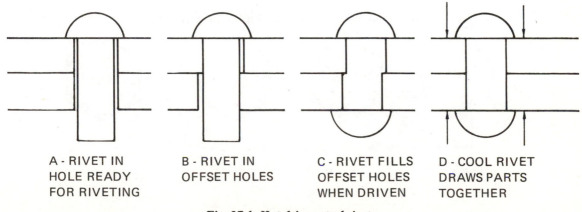

A - RIVET IN
HOLE READY
FOR RIVETING

B - RIVET IN
OFFSET HOLES

C - RIVET FILLS
OFFSET HOLES
WHEN DRIVEN

D - COOL RIVET
DRAWS PARTS
TOGETHER

Fig. 27-1 Hot-driven steel rivets

261

in an expanded state when the second head is formed. Upon cooling, the rivet contracts and draws the steel parts more tightly together, figure 27-1d, page 261.

BOLTS

There are two general classes of structural bolts, common bolts and high-strength bolts. *Common bolts* are made of low-carbon steel and can be identified by their square heads. Common bolts are used for certain classes of structures not subject to high loads. They also are used for fitting up structures which are to be welded. They are less expensive than high-strength bolts.

Because common bolts fit loosely in the holes and cannot be tightened enough for significant frictional resistance, their capacity is reduced. The allowable load on common bolts, is 3/4 of that allowed for rivets of the same size.

HIGH-STRENGTH BOLTS

High-strength structural bolts satisfy ASTM Specifications A325 and A490. These bolts are made of special steels which are so strong that they can be tightened enough to firmly clamp the pieces together. When they are tightened properly, the friction between the parts carries the load; the bolt does not need to fit the holes exactly when this is done. A load equal to that of hot-driven rivets of the same size is allowed for the A325 bolts. The A490 bolt has even greater capacity than the A325 bolt.

The advantages of using high-strength bolts instead of rivets are several. Installation is relatively noiseless compared with riveting. In the field, bolts can be installed by a two-man crew instead of the four men required in a riveting gang. High-strength bolts give a connection with much greater resistance to vibration and stress reversal. High-strength bolts used in fitting up a structure can be left in place after final alignment.

IDENTIFICATION OF HIGH-STRENGTH BOLTS

High-strength bolts have special identification. A325 bolts are made in three types. The mechanical properties of the three types are comparable. Type I is automatically furnished unless otherwise specified. Type II is made of slightly different steel. Type III has weathering characteristics and would be specified for use with the new weathering steels. The characteristic markings for these three types of A325 bolts are shown in figure 27-2. A490 bolts require washers under the head and nut when used with A36 steel members. For use with higher-strength steels, a washer only under the nut is required.

BOLT SYMBOLS AND NOTE ON PLANS

Bolts are indicated on plans by a note. Details of connections show field bolts

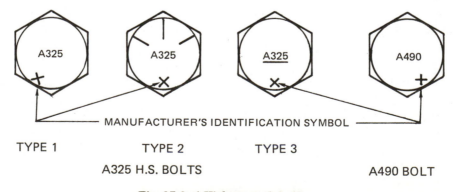

MANUFACTURER'S IDENTIFICATION SYMBOL

TYPE 1   TYPE 2   TYPE 3

A325 H.S. BOLTS   A490 BOLT

Fig. 27-2 4 High-strength bolts

indicated by solid round dots with shop bolts or rivets indicated by open circles, figure 27-3. Sometimes bolts are indicated simply by a center line drawn through the parts, figure 27-4.

WELDING

Arc welding is the fusion of metal by an electric arc. Arc welding has been developed to the point where it is now the most important fastening method for structural steel. Early problems with brittle welds are solved by the shielded-arc welding process. Field welding is done by hand using an electrode with a coating that provides the gaseous shield, figure 27-5, page 264. The shield keeps oxygen and nitrogen from being picked up by the molten metal. The ductility of the weld is maintained and thus dependable welds with allowable capacities are possible. Shop welding by automatic machines usually is done by the submerged-arc process, figure 27-6, page 264, using bare-wire electrodes.

Specifications for electrodes and standards of good welding practice have been adopted by ASTM and the American Welding Society (AWS). Electrodes commonly used in structural work are E60 and E70 series. The capacity of the weld also depends upon the base metal (the steel being welded). Handbooks on welding give detailed descriptions of electrode types and the kinds of work for which they are suited.

Welds are described by the type of joint on which they are used. If the space or groove

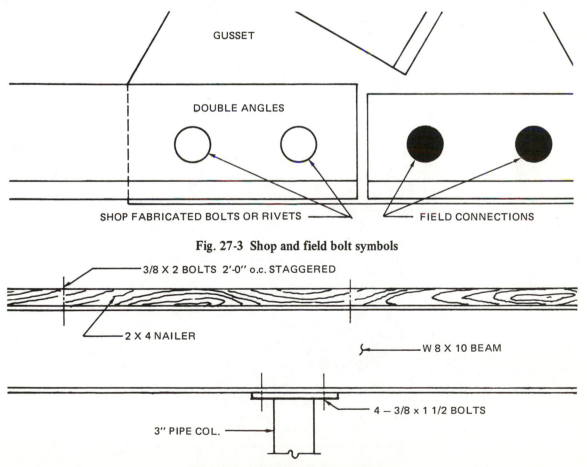

Fig. 27-3 Shop and field bolt symbols

Fig. 27-4 Bolts indicated by center lines

between two parts is filled with weld metal, it is a *groove* or *butt weld*.   A *fillet weld* is one placed in a corner forming a triangular-shaped weld.   The fillet weld is by far the most used weld when connecting structural steel building members.   Various types of welds are shown in figure 27-7.

The size of a fillet weld is measured along the base of the triangular fillet, figure 27-8.   The strength is based upon the throat

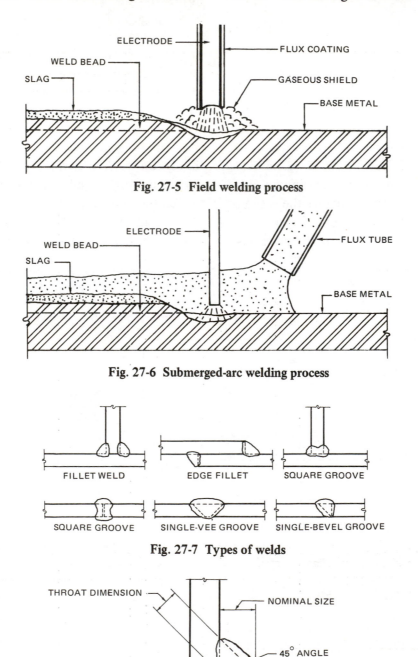

Fig. 27-5 Field welding process

Fig. 27-6 Submerged-arc welding process

Fig. 27-7 Types of welds

Fig. 27-8 Size of fillet weld

dimension regardless of the direction of the load. The maximum-sized fillet weld that can be made with one pass is 5/16 of an inch. Larger welds require multiple passes and are avoided if possible because of the expense. The size of the weld, as well as other important information, is given on the weld symbol, figure 27-9.

WELDED JOINTS
Standard symbols

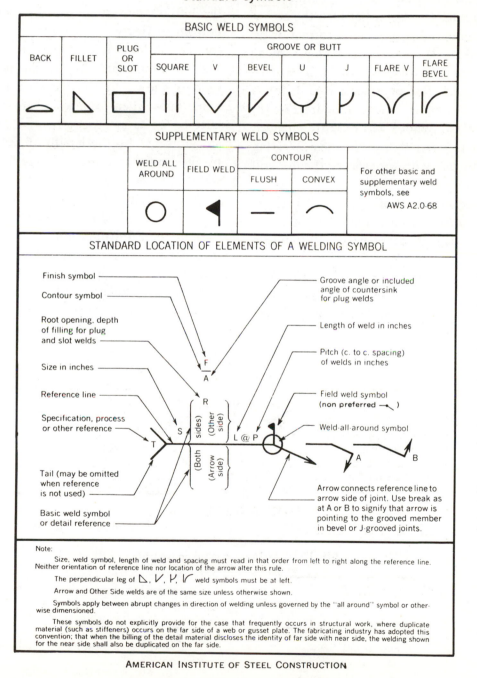

Fig. 27-9 Welded joints, standard symbols

WELDING SYMBOLS

The convention used to locate the welding information on drawings is a horizontal reference line with a sloping arrow connecting it to the joint, figure 27-10. The arrow may be pointing right or left, upward or downward, but always at an angle to the reference line. The connecting point between the arrow and the reference line provides information as in figure 27-11. If no extra marking is shown, a shop weld is indicated, figure 27-11a. When a small triangular flag or a solid round dot is shown a field weld is indicated, figure 27-11b. The field weld symbol was recently changed from a dot to a flag by the American Welding Society, but many drawings using the dot still exist. An open circle means weld all around the member in the shop, figure 27-11c. If the open circle is shown around the dot or the base of the flag, it means weld all around in the field, figure 21-11d.

The basic weld symbol is located about midway on the horizontal reference line, figure 27-12. The symbol is below the line if the weld is to be placed on the near side where the arrow points, figure 27-12a. It is placed above the line if the weld is to be placed on the far side, figure 27-12b, and above and below if both sides are to be welded, figure 27-12c. The perpendicular leg of certain basic symbols always is drawn on the left of the symbol (according to note on bottom of figure 27-9). The size of the weld (or its depth) is indicated to the left of the basic symbol, figure 27-13. The length of the weld is shown to the right of the symbol. Sometimes when long joints are made, intermittent welds are used, figure 27-14. These are indicated by the length of weld followed by the center to center spacing (pitch). Such welds may be staggered on either side of the joints as in figure 27-14b.

At the tail of the reference line may be located some information about the kind of material or process required, figure 27-15a. This feature is not often used in structural steel details. When no information is needed, the tail is left off, figure 27-15b.

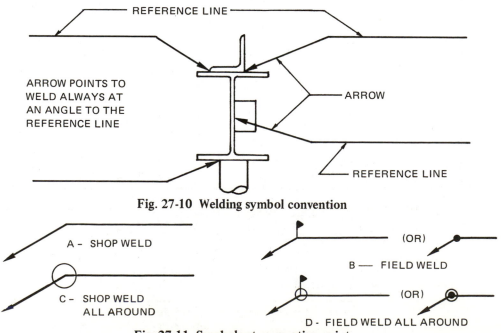

Fig. 27-10 Welding symbol convention

Fig. 27-11 Symbols at connecting point

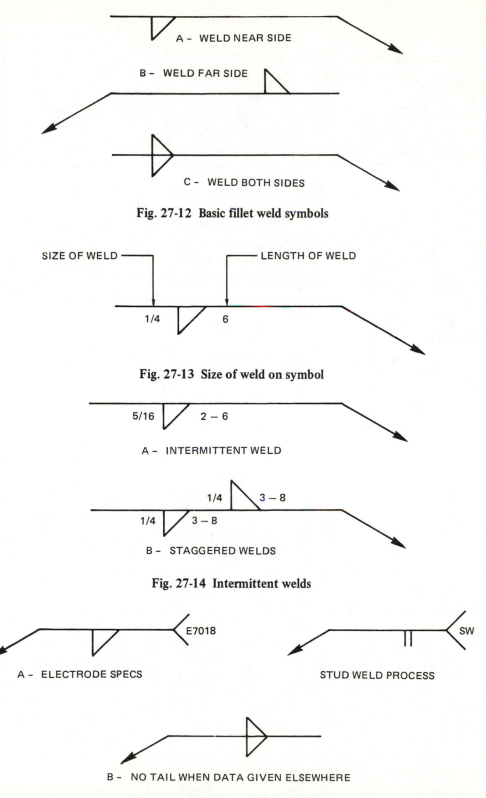

A – WELD NEAR SIDE

B – WELD FAR SIDE

C – WELD BOTH SIDES

Fig. 27-12 Basic fillet weld symbols

SIZE OF WELD ——⌐   ⌐—— LENGTH OF WELD

1/4   6

Fig. 27-13 Size of weld on symbol

5/16   2 – 6

A – INTERMITTENT WELD

1/4   3 – 8

1/4   3 – 8

B – STAGGERED WELDS

Fig. 27-14 Intermittent welds

E7018   SW

A – ELECTRODE SPECS   STUD WELD PROCESS

B – NO TAIL WHEN DATA GIVEN ELSEWHERE

Fig. 27-15 Welding symbol tail data

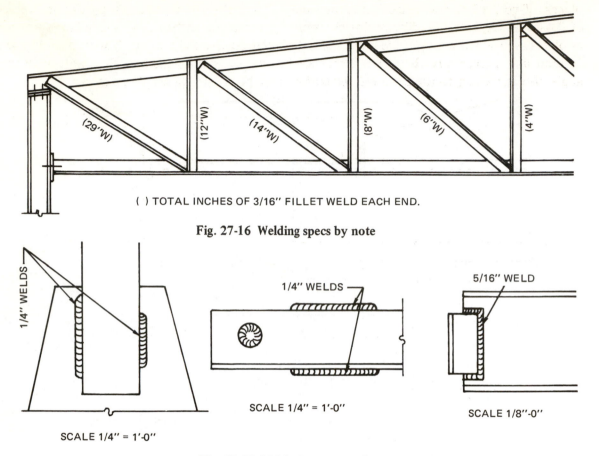

( ) TOTAL INCHES OF 3/16" FILLET WELD EACH END.

Fig. 27-16 Welding specs by note

1/4" WELDS

SCALE 1/4" = 1'-0"

1/4" WELDS

SCALE 1/4" = 1'-0"

5/16" WELD

SCALE 1/8"-0"

Fig. 27-17 Welds drawn to scale

In addition to these standardized ways of indicating welds, some plans use a simple note as in figure 27-16. In this case all welds are the same size and the lengths are indicated for each joint. Another variation is to draw the weld, either solid or outlined, in its proper place and label the size, figure 27-17. If a note points to a joint and says weld, the size is not critical. A minimum-sized weld based upon the thicknesses of the parts should be used.

EXAMPLES

The following examples of welding symbols are given to help in the understanding of their use. They are the ones typically found on construction drawings:

Example 1. Double angles are welded to a gusset plate as in a steel truss. The welding is to be done in the shop. The welds are 3/16" fillet welds on both sides of the joint. The perpendicular side of the fillet triangle is to the left in each case, figure 27-18.

Example 2. A pipe column cap plate is shown welded in the shop all around with a 1/4" fillet in figure 27-19a. A base detail is shown in figure 27-19b. Both sides of the anchor are welded in the shop to the bottom of the base plate. The base plate is set in concrete and later the column is field welded all around with a 1/4" fillet weld.

Example 3. The girder to column connection in figure 27-20 is prepared in the shop in three places. Two 3 x 3 angles are welded to the web of the girder (A) with 1/4″ fillet welds. Stiffener plates are welded between the flanges of the column in a similar way (B). A 6 x 4 erection angle also is welded to the column. After the girder is put in place, its flanges are welded to the column with 3/4″ single bevel butt welds. These are field welds. The outstanding legs of the web angles also are field welded all around and on both sides of the girder.

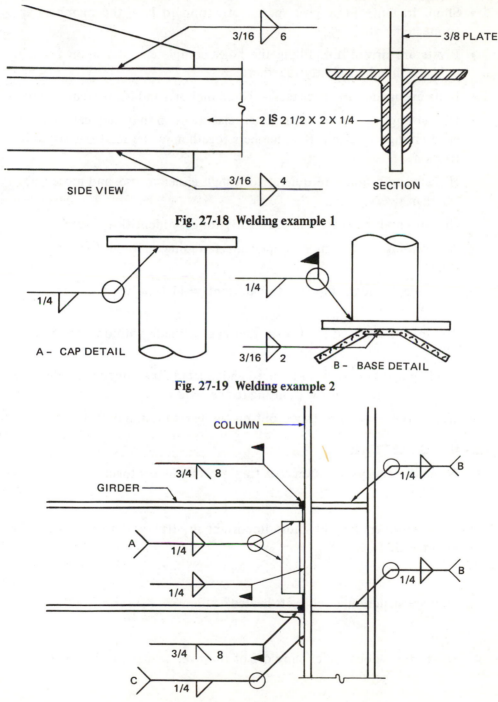

Fig. 27-18 Welding example 1

Fig. 27-19 Welding example 2

Fig. 27-20 Welding example 3

SUMMARY

- Steel members are selected to safely carry their loads on the basis of their properties.

- Transferring loads to other members is the job of the connections.

- Shop drawings must give special attention to how the members are connected.

- Rivets are driven hot, filling the holes completely, and upon cooling, hold the members tightly together.

- Bolts are made in two classes — common bolts and high-strength bolts.

- High-strength bolts are made of special steels so that they can be tightened enough to clamp the members together so the load is resisted by friction.

- High-strength bolts are quieter to install, easier to use, and more versatile than rivets.

- High-strength bolts are marked clearly for easy identification.

- Arc-welding is the most important fastening method for structural steel.

- The fillet weld is the most important weld type used on structural steel members.

- The maximum-sized fillet weld that can be made with one pass is 5/16 of an inch.

- Welding symbols are standardized and located on a horizontal reference line with a sloping arrow pointing to the weld.

- Welds sometimes are drawn out on the parts or simply noted "weld."

REVIEW QUESTIONS

1. Name three ways to fasten structural steel members together.

2. What fastener was the most important in structural steel work until about 1950?

3. Why are structural steel rivets heated?

4. What effect does the cooling off of the rivet have upon the parts being fastened together?

5. Give one of the important reasons why rivets are not used for field erection of structural steel today.

6. Common steel bolts, primarily used for fitting up structures for welding, can carry only _____ of the load a rivet the same size could support.
   a. 1/2.   c. 3/4.
   b. 3/8.   d. 7/8.

7. When properly tightened, what causes high-strength bolts to carry the load between parts?

8. High-strength A325 bolts have a load capacity of _____ times that of a rivet of the same size.
   a. 3/4.   c. 1 1/4.
   b. 1.   d. 2.

9. In detail drawings of steel connections, each bolt or rivet is indicated. Sketch the indication for bolts or rivets when:
   a. shop assembled.

   b. field assembled.

10. Which of the following welding methods is used for field welding?
    a. shielded arc using coated electrodes.
    b. submerged arc using bare-wire electrodes.

11. Select the most important type of weld used to connect structural steel.
    a. plug.   c. intermittent.
    b. groove.   d. fillet.

12. How are high-strength bolts marked to identify them?

13. Complete the weld symbol shown if the welds are 1/4 inch fillets on both sides of the part, 3 inches in length, and to be field welded using E70 electrodes.

14. Sketch a weld symbol indicating an intermittent fillet weld, 1/4 inch in size, 2 inches long, spaced 6 inches on center, and staggered on each side of the part.

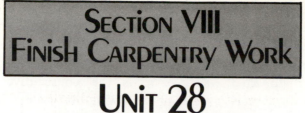

SECTION VIII
Finish Carpentry Work

Unit 28
Interior Finish

OBJECTIVES

After studying this unit, the student should be able to:

- explain how doors are identified on the drawings.
- tell what is included in a hardware schedule.
- describe two types of grid ceiling systems.
- sketch various types of stock moldings.

DOORS — TYPES AND SCHEDULES

Doors and their frames, hardware, and installation details must be understood by the finish carpenter. Architectural drafting conventions are used on the floor plans to include the location, type, and swing of doors. Figure 28-1 shows several common types of doors. The size of the door may be indicated at the door opening, figure 28-2. On most commercial projects, the doors and frames are

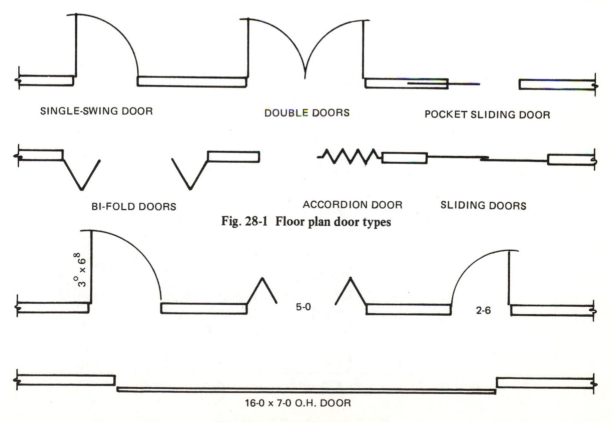

Fig. 28-1 Floor plan door types

Fig. 28-2 Door size on floor plan

identified in a schedule. Several systems of numbering the door openings are used. The door number usually is placed in a circle or other distinctive shape. Examples of door numbering are shown in figure 28-3.

Perhaps the simplest system is to assign a number to each different size of door. The identification is given in a schedule or by sketches, figure 28-4. By this method, every 3'-0" x 6'-8" door is given the same code number.

A more exact system numbers every door opening in consecutive order. This makes checking the door schedule more sure and is an excellent way to provide other information. An example of this type of schedule is given in figure 28-5.

A third method is to include two numbers in the door symbol, figure 28-3c. The top number is the room number which the opening serves. The bottom number is the consecutive number of the doors serving that room. If there is only one door, the number is one. If there are four doors, there is a one, a two, a three, and a four on the various door notations. This method enables the reader to know which room is served since the room number is part of the symbol.

Although architects use different systems, there are similarities. A brief study of the plans and schedules usually is sufficient to understand their meaning.

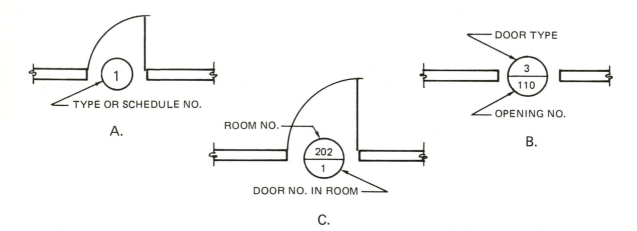

Fig. 28-3 Door schedule numbering systems

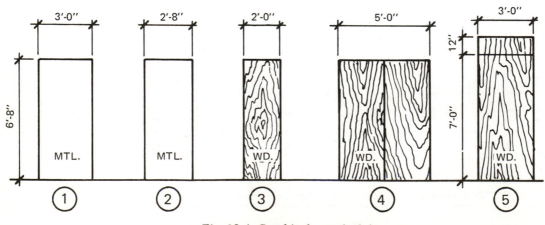

Fig. 28-4 Graphic door schedule

HARDWARE

Hardware is a bit more complicated. Groups of hardware sometimes are identified in the door schedule, figure 28-5. Knowing the group number, the list of hardware items for an opening is found in the specifications,

figure 28-6. Sometimes a hardware schedule is given that lists each opening with the required items, figure 28-7.

Most contract hardware is listed by the vender in a detailed hardware schedule that is submitted like shop drawings for approval.

DOOR SCHEDULE					
NO.	SIZE	THICK.	FRAME & TYPE		HARDWARE GROUP
1	3 - 0 X 6 - 8	1 3/4	MTL.	2	2
2	2 - 9 X 6 - 8	1 3/4	MTL.	2	2
3	2 - 0 X 6 - 8	1 3/8	MTL.	3	1
4	2 - 0 X 6 - 8	1 3/8	MTL.	4	4
17	3 - 0 X 7 - 0	1 3/8	MTL.	3	1
18	3 - 0 X 7 - 0	1 3/4	MTL.	2	2
19	3 - 0 X 7 - 0	1 3/4	MTL.	2	5

Fig. 28-5 Schedule listing every opening

GROUP 1

BUTTS 3 1/2 X 3 1/2
LOCK D80 PLY 28
STOP 331 ES

GROUP 2

BUTTS 1146 5 X 41/2
LOCK D80 PLY 28
STOP 331 ES

GROUP 5

BUTTS BB 1146 5 X 4 1/2
LOCK D80 3/4 PLY 28
CLOSER D HFL
ARMOR PLATE 48 32D
STOP 331ES
2 EDGINGS D 32D

Fig. 28-6 Hardware groups from specs

SYMBOL	DESCRIPTION		LOCK A10S NOVO X 10STMS	LOCK A52PD NOVO X 10SMI	1 1/2 PR BUTTS 174A5 41/2 X 41/2	1 1/2 PR BUTTS BB 241 G25 X 5	DOOR CLOSER C REG. MS	DOOR STOP FB 13 X 5US10	DOOR HOLDER 1150 US10		
	LOCATION	HAND									
1	HALL TO STAIRS 3 - 0 X 6 - 8 X 1 3/4	RHRB	●		●		●				
2	HALL TO CLASSRM 3 - 0 X 6 - 8 X 1 3/4	LHRB		●	●			●	●		

HARDWARE SCHEDULE

Fig. 28-7 Graphic hardware schedule

Each and every opening is listed, followed by the hardware to be furnished. Catalog numbers, finishes, and quantities are listed carefully, figure 28-8. The locksets are identified from this schedule when they are shipped to the job. They are handed for the opening they are to be used in. This is important to the person installing the hardware. It saves time by not having to set the hand of the lockset. The *hand* of a door means the way it swings. Figure 28-9 shows the standards used for classifying door openings.

Steel door frames are furnished with hardware preparation included. It is imperative that the carpenter set the proper frame in each opening so the door and hardware installed later will be correct. Steel door frame handing is somewhat simpler than for locksets. Only right-hand or left-hand frames are scheduled, figure 28-10. In identifying steel

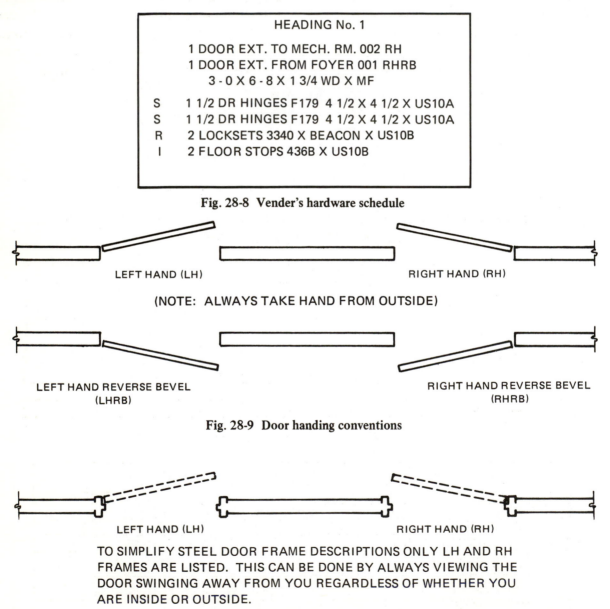

HEADING No. 1

1 DOOR EXT. TO MECH. RM. 002 RH
1 DOOR EXT. FROM FOYER 001 RHRB
3 - 0 X 6 - 8 X 1 3/4 WD X MF

S   1 1/2 DR HINGES F179   4 1/2 X 4 1/2 X US10A
S   1 1/2 DR HINGES F179   4 1/2 X 4 1/2 X US10A
R   2 LOCKSETS 3340 X BEACON X US10B
I   2 FLOOR STOPS 436B X US10B

Fig. 28-8 Vender's hardware schedule

LEFT HAND (LH)   RIGHT HAND (RH)

(NOTE: ALWAYS TAKE HAND FROM OUTSIDE)

LEFT HAND REVERSE BEVEL   RIGHT HAND REVERSE BEVEL
(LHRB)   (RHRB)

Fig. 28-9 Door handing conventions

LEFT HAND (LH)   RIGHT HAND (RH)

TO SIMPLIFY STEEL DOOR FRAME DESCRIPTIONS ONLY LH AND RH FRAMES ARE LISTED. THIS CAN BE DONE BY ALWAYS VIEWING THE DOOR SWINGING AWAY FROM YOU REGARDLESS OF WHETHER YOU ARE INSIDE OR OUTSIDE.

Fig. 28-10 Steel frame handing conventions

door frames, the hand always is taken with the door swinging away from the viewer.

ROOM NUMBERS

Floor plans have each room and space numbered as well as named, figure 28-11. The room number is used in the various schedules for reference purposes. A common system of numbering uses the floor level as the first digit. First floor rooms are numbered 100, 101, 102, etc., usually arranged in a counterclockwise sequence around the plan. Ground floors or basements begin with room number 1 and continue as required. This system helps the reader to know on which floor of the building a room is located. For emphasis, the room number is often boxed as in figure 28-11.

ROOM FINISH SCHEDULES

The materials and finishes of exposed interior room surfaces usually are given in a finish schedule. This may be contained in the drawings or in the specifications. Figure 28-12 is an example taken from specifications. Figure 28-13 is a material and finish key to

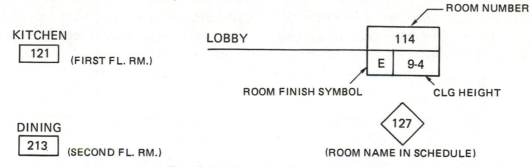

Fig. 28-11 Floor plan room numbers

NO.	TITLE	FLOOR		BASE		WALLS		CEILING			NOTES
		M	F	M	F	M	F	M	F	HT	
101	LOBBY	A	4			A	1	L	1	9-6	
102	VESTIBULE	A	4			A	1	L	1	9-6	
108	JANITORS CLO.	H	3	Q	1	B	2	L	1	8-0	
113	COAT ROOM	K	1	Q	1	B	2	L	1	9-6	
114	STORAGE	H	3	Q	1	B	2	L	1	9-6	
116	MULTI-PURPOSE	K	1	Q	1	F	1	L	1	9-6	
122	RECEIVING	J	1	J	1	B	2	L	1	8-0	

Fig. 28-12 Finish schedule from specs

MATERIAL KEY		FINISH KEY	
MK	DESCRIPTION	MK.	DESCRIPTION
A	BRICK	1	CLEANED
B	CONCRETE BLOCK	2	PAINTED
H	VINYL ASBESTOS TILE	3	CLEAN & WAX
J	SEAMLESS COVERING	4	CLEAN & SEALED
K	CARPET N.I.C.	5	STAIN & VARNISH
L	LAY-IN ACOUS TILE	6	SANDBLASTED
Q	VINYL BASE		

Fig. 28-13 Key to finish schedule

explain the schedule. Room 113, the Coat Room, for instance has carpet (K) on the floor, a vinyl base (Q), concrete block (B) walls, and a lay-in accoustical ceiling (L) 9 feet 6 inches high. The room number 113 denotes a first floor room.

Some schedules give the finish directly in the schedule instead of using a key, figure 28-14. This makes the schedule a little bit easier to read but takes up more space on the drawing. Space usually is made available in all schedules for notes or remarks.

PARTITIONS

Partitions are interior walls used to divide a building into smaller areas or rooms. Two general types are used, permanent and movable. Permanent partitions constructed by the carpenter are made with wood or with lightweight metal framing, figure 28-15. Surface materials applied include plywood, paneling, gypsum board, and plastic panels. The carpenter must be able to lay out the partitions as shown on the plans. The wall symbols usually indicate the materials to be used. Sections give more information and the specifications complete the instructions.

Movable partitions are manufactured and shipped to the job and require setting only. However, approved shop drawings should be followed. Details vary with different manufacturers. Very simple when viewed on the plans, these walls sometimes are complicated in their joints and details. Typical sections are shown in figure 28-16.

MK.	FLOOR	BASE	WAINSCOT	WALLS	CEILING
A	CONCRETE	NONE	NONE	LT. WT BLK PAINT	LAY-IN ACOUS TILE
B	VINYL TILE	VINYL	DO	DO	DO
C	VINYL TILE	VINYL	DO	GYP. BD. PAINT	DO
D	CARPET	NONE	DO	PRE-FIN PANEL	DO
E	CARPET	WOOD	PANELING	WALLPAPER	GYP. BD. PAINT
F	CER. TILE	CER. TILE	CER. TILE	LT. WT. BLK PAINT	LAY-IN MTL. TILE
G	BRICK	NONE	NONE	BRICK	LAY-IN ACOUS TILE

Fig. 28-14 Complete finish schedule

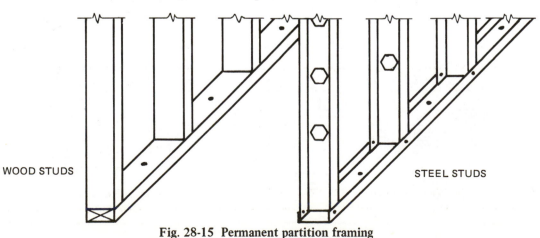

WOOD STUDS   STEEL STUDS

Fig. 28-15 Permanent partition framing

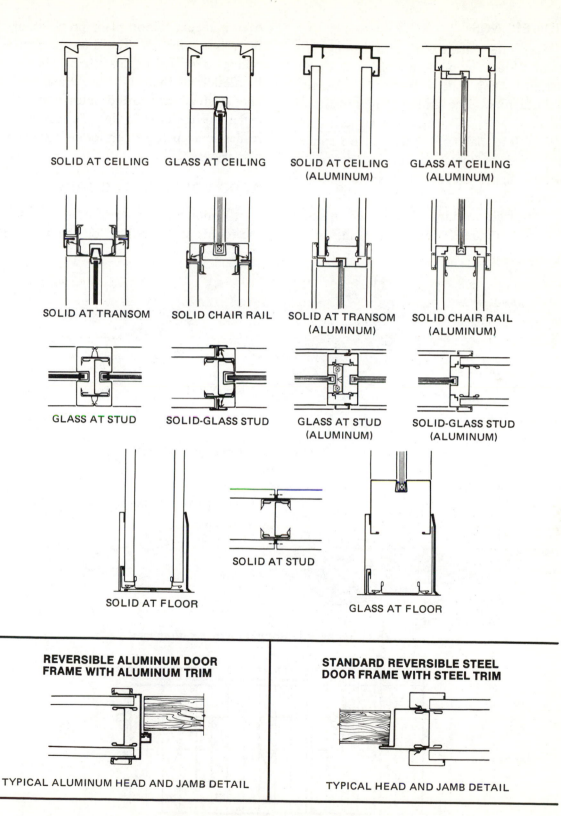

SOLID AT CEILING   GLASS AT CEILING   SOLID AT CEILING (ALUMINUM)   GLASS AT CEILING (ALUMINUM)

SOLID AT TRANSOM   SOLID CHAIR RAIL   SOLID AT TRANSOM (ALUMINUM)   SOLID CHAIR RAIL (ALUMINUM)

GLASS AT STUD   SOLID-GLASS STUD   GLASS AT STUD (ALUMINUM)   SOLID-GLASS STUD (ALUMINUM)

SOLID AT STUD

SOLID AT FLOOR   GLASS AT FLOOR

REVERSIBLE ALUMINUM DOOR FRAME WITH ALUMINUM TRIM

TYPICAL ALUMINUM HEAD AND JAMB DETAIL

STANDARD REVERSIBLE STEEL DOOR FRAME WITH STEEL TRIM

TYPICAL HEAD AND JAMB DETAIL

Fig. 28-16 Movable partition sections

GRID CEILINGS

Suspended acoustical grid ceilings are very important in modern commercial construction. They usually are set by specialty subcontractors. The grid system consists of main tees, cross tees, and wall angles, figure 28-17. More elaborate systems have splines (thin wood or metal strips) and concealed supports, figure 28-18. The specs give the basic requirements.

A reflected ceiling plan is used to show the layout, figure 28-19. This plan, drawn over a regular floor plan, pretends that the floor is a mirror. The drawing depicts the ceiling as viewed in this mirror. In this way, the details of the ceiling are drawn with solid lines instead of dotted lines. It is much easier to understand in this way since the reader is accustomed to looking down on the plan views.

WOOD MOLDINGS AND TRIM

Running trim consists of moldings around the rooms at the ceiling, on the walls, and at

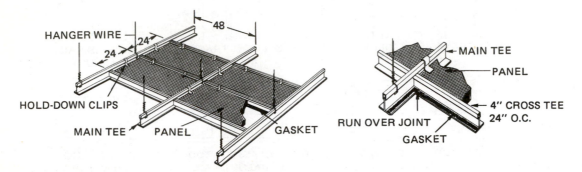

Fig. 28-17 Suspended ceiling exposed grid system

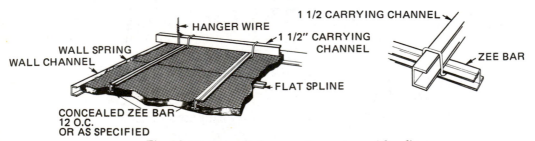

Fig. 28-18 Concealed suspension system with splines

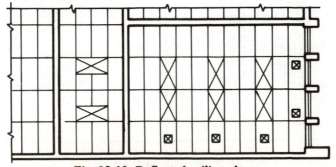

Fig. 28-19 Reflected ceiling plan

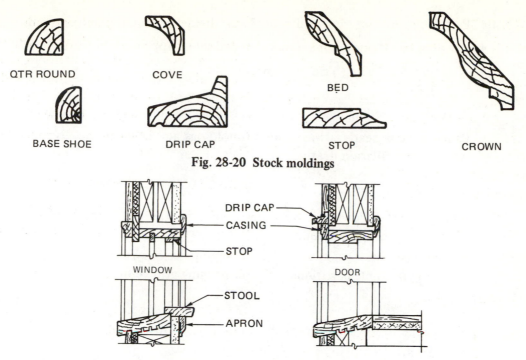

Fig. 28-20 Stock moldings

Fig. 28-21 Door and window details

the floor. Door and window trim is detailed to be compatible. Typical stock moldings are shown in figure 28-20. The detail wall sections usually indicate this work. The specifications also should be consulted. Door and window setting details clearly show the trim required, figure 28-21.

Paneled walls nearly always require moldings at certain points. Sometimes the carpenter must add moldings at some joints to complete the installation neatly. The plans do not always show every piece required for a complete job.

Skill and knowledge of many people are needed to finish the inside of a building. The finish carpenter is one such person. The quality of the building depends much upon the performance of the builder.

SUMMARY

- The carpenter may set metal door frames, hang doors, build and finish partitions, install grid ceilings and movable partitions.
- On most commercial type projects, the doors and frames are identified in a schedule in the plans.
- Hardware is listed by the vender in a detailed hardware schedule that is submitted like shop drawings for approval.
- The hand of a door means the way it swings.
- Steel door frames are furnished either right-handed or left-handed.
- Every room and space is named and numbered on architectural floor plans.
- The materials and finishes of interior room surfaces usually are given in a finish schedule.

- Partitions are interior walls used to divide a building into rooms.
- Movable partitions are manufactured and set by approved shop drawings.
- A suspended ceiling grid system consists of main tees, cross tees, and wall angles.
- A reflected ceiling plan is used to show the layout of the ceiling.
- Paneled walls nearly always require moldings at certain points to complete the installation neatly.

REVIEW QUESTIONS

1. Which of the following is work never done by the carpenter when finishing the interior of a building?

   a. hanging doors
   b. setting movable partitions

   c. laying masonry partitions
   d. hanging grid ceilings

2. Sketch the following doors as shown on floor plans.

   a. single swing door

   b. pocket sliding door

   c. bi-fold doors

3. On drawings, what is the purpose of the number usually found in a circle at each door opening?

4. Explain the meaning of the numbers in this door marking.

    a.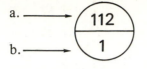

    b.

5. When hardware group numbers are included on the door schedule, where is the list of required items for the groups explained?

6. The hand of a door means the way it swings. Identify the hand of the following.

    a.

    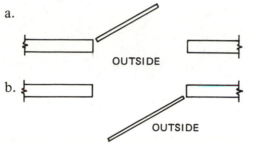

    OUTSIDE

    b.

    OUTSIDE

7. Steel door frames are listed as either LH or RH. How is this possible?

8. Rooms are numbered on floor plans for easy reference. How is the floor level indicated in these numbers?

9. Where are the room finish schedules found?

10. What type partitions require shop drawings because they are manufactured away from the job site?

11. Name the three components of a typical suspended ceiling grid.

12. What is a reflected ceiling plan?

13. Sketch the following stock moldings.
     a. quarter round

     b. cove

     c. bed

14. What is needed to complete the installation of most prefinished paneling
     jobs?

Unit 29
Cabinets, Fixtures, and Stairs

OBJECTIVES

After studying this unit, the student should be able to:

- explain how cabinets are indicated on the floor plans.
- tell how standardized cabinets are detailed so they may be installed between two walls.
- give two rules used to proportion treads and risers to maintain equal strides for stairs.

SPECIAL MILLWORK

Cabinets, fixtures, and stairs are high-class millwork. They are rarely built on the job in commercial carpentry practice. Instead, they are made in well-equipped cabinet shops and woodworking mills. Therefore, the carpenter's main task is the setting of this special millwork.

KITCHEN CABINETS

Kitchen cabinets are the most common type of cabinets installed by the carpenter. The conventions and details used to show them are the same as other cabinet work, such as wardrobes, linen closets, and utility room cabinets. Cabinets are shown on floor plans by simple outlines drawn to scale, figure 29-1.

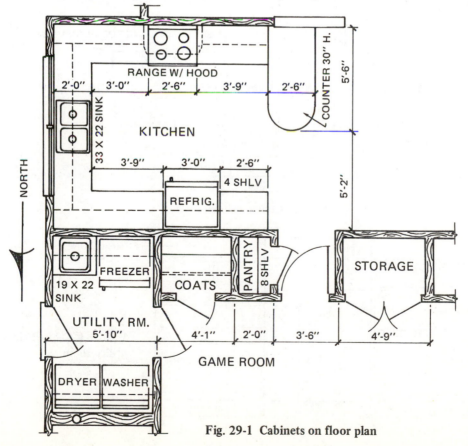

Fig. 29-1 Cabinets on floor plan

285

The base cabinets, when 36 inches high, are below the section plane of the plan view. They are, therefore, indicated by solid lines. The wall cabinets, being above the plane of the plan view, are shown as dotted lines. A typical section of a kitchen cabinet is shown in figure 29-2. A line representing the plan view section plane is indicated there.

The typical section, figure 29-2, gives the dimensions and construction details at one certain place in the cabinet. In order to give all the information needed to install the cabinets, elevation views are used. These elevation views are usually made of the entire wall where the cabinets are located, figure 29-3.

All the doors, drawers, shelves, and appliances on that wall are shown and dimensioned. Door swings are indicated by dotted lines angling from the center of the hinged side to the opposite upper and lower corners. When a pair of doors is used, the swing symbol forms a diamond shape. This same swing symbol is used for casement windows on elevations.

When the cabinet continues around the corner of the room, the elevation view shows this. The portion of the cabinet cut by the elevation plane is usually shown by an outline only, figure 29-4. Section details can be included, but this requires considerable extra

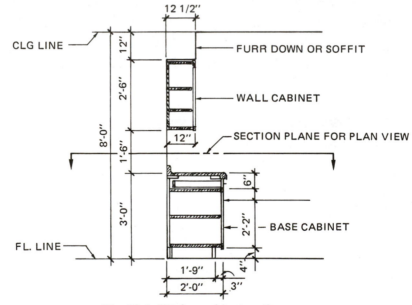

Fig. 29-2 Kitchen cabinet section

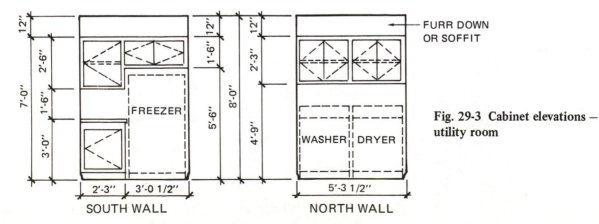

Fig. 29-3 Cabinet elevations — utility room

drafting time. Usually, no new information is given that cannot be found in the typical section and on the elevations.

When the cabinets are located on two or more walls in the room, the elevations must be identified. Two common methods used to identify them are by compass direction and by a key plan with symbols. When the orientation of the building is indicated by a north arrow, the walls are named for their compass direction. This works well when the plan is not too large or complicated. For large buildings with many different rooms containing cabinets, key plans drawn of each room are used, figure 29-5. A symbol on the plan points to a wall with cabinets. The elevation of that wall is labeled with the same symbol.

Shop drawings are usually required when the cabinets are made away from the job. Field measurements are made after the walls are framed and finished. These measurements are necessary to verify the dimension of the wall spaces where the cabinets are installed. Extra material is then detailed into the cabinets so they can be trimmed to fit exactly to the walls.

Standardized cabinets are sometimes used. These cabinets consist of many smaller units that are combined to provide the functional requirements shown on the architect's plans, figure 29-6, page 288. Standardized cabinets are made in sizes based upon multiples of 3 inches. By selecting a proper combination of these units, the space between

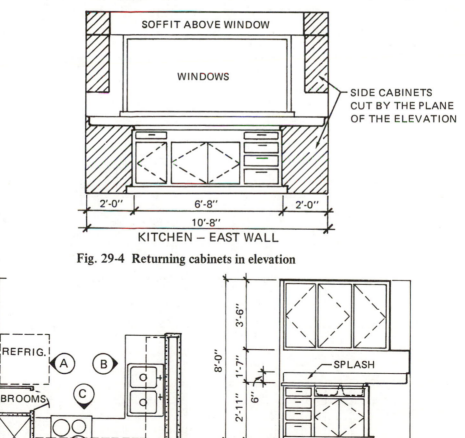

Fig. 29-4 Returning cabinets in elevation

Fig. 29-5 Key plan identifies cabinet elevations

walls is filled. A small clearance is required to install the units. This clearance space is covered with filler strips of 1, 2, or 3 inches. If necessary, these filler strips can be placed at each end of the assembled units. Wider fillers are not generally used because a three inch wider cabinet unit can be selected somewhere to lengthen the layout.

Shop drawings for standardized cabinets are usually quite simple. Each unit has a catalog number which includes the width of the unit. Once the distance between walls is known, the units and fillers are selected and noted, figure 29-7. The combined width of the units in a cabinet is checked by adding the unit numbers. The cabinet top is made to

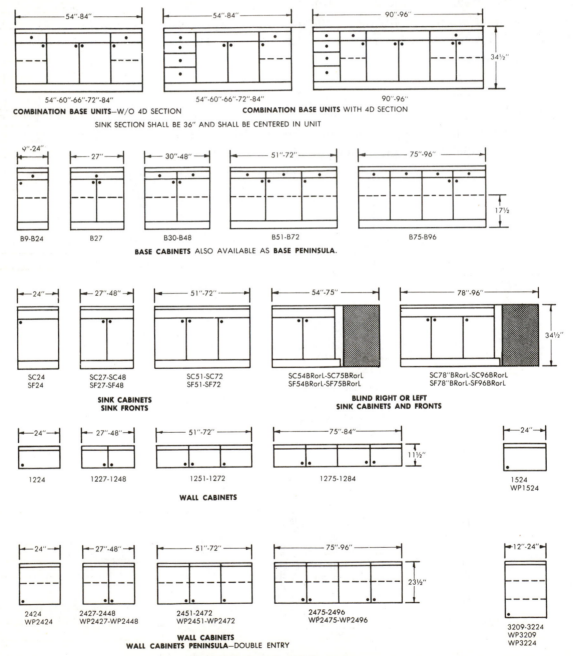

Fig. 29-6 Standardized cabinets

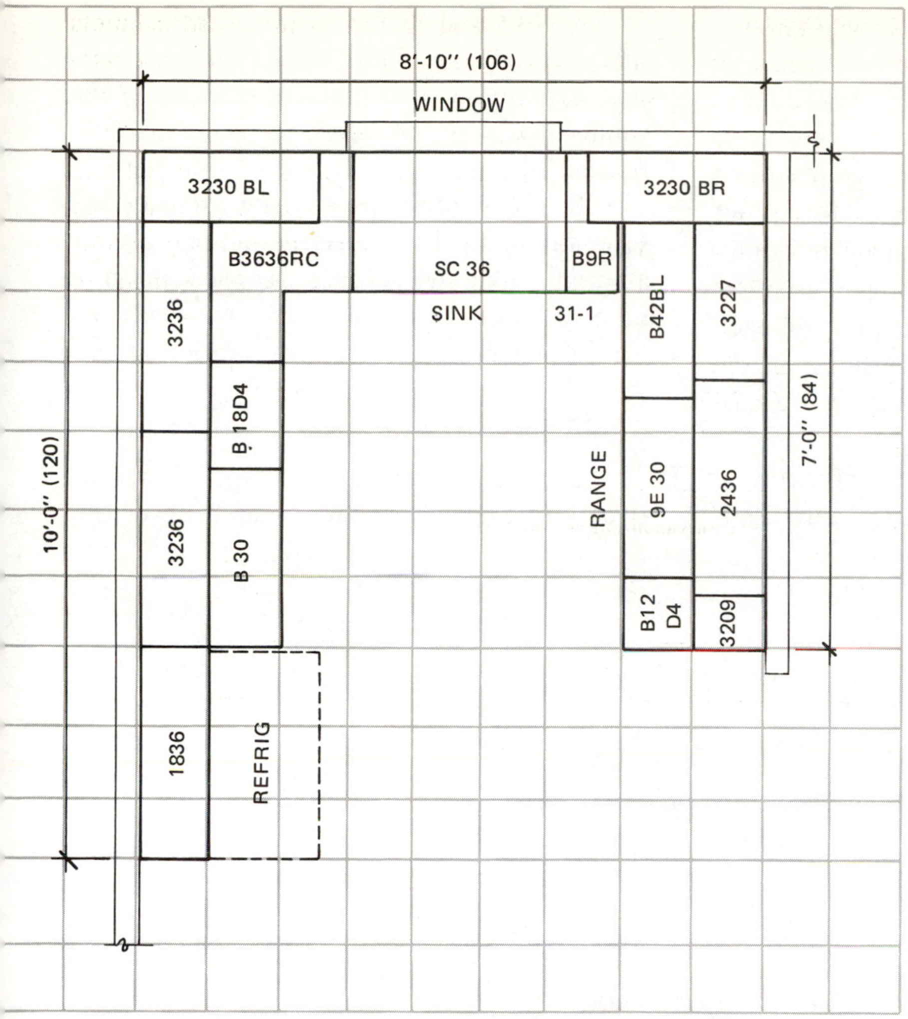

BILL OF MATERIALS

SINK & BASE CABINETS

Quan.	Model	Price	Ext.
1	SC36		
1	B42BL		
1	B3636RC		
1	B 30		
1	B 12 D4		
1	B 18 D4		
1	B9R		

OVEN-RANGE & UTILITY CABS

Quan.	Model	Price	Ext.
1	GE30		

WALL CABINETS

Quan.	Model	Price	Ext.
1	3230 BR		
1	3230 BL		
2	3236		
1	3227		
1	3209		
1	1836		
1	2436		

BATH CABINETS

FBP & MISC.

WALL SCRIBES (WS)

Quan.	Model	Price	Ext.
1	31-1		

Finish **CONTINENTAL**

Type Wood **BIRCH**

Custom Line

Fairview Line ✓

Type Doors:

Upper **FLUSH**

Lower "

Type Drawers:

Fronts & W/Panels

Regular Wood ✓

Wood - W/KV 1300 Guide

Type Pulls

Drill

Drill Base Only

Do Not Drill ✓

Base Backs

Cut Outs: (size)

Oven

Fig. 29-7 Shop drawings for standardized cabinets

cover the several units and would be the same as in custom cabinet construction.

STAIRS

Stair building is another type of special finish carpentry work. The stair design and details are carefully worked out by the architect. The floor plans indicate the location, number of risers, and traffic flow of each flight, figure 29-8. The dimensions of the stair enclosure and its construction are also indicated on the floor plans.

Various arrangements of stair flights and landings are found in buildings, figure 29-9.

Sometimes spiral stairs are used where limited space is available, figure 29-10. Another design used in limited space is the stair using winders, instead of landings at the turns, figure 29-11.

In fire-resistive construction, stair enclosures are required by code in order to evacuate a burning building. Additional open stairs may be used in special cases. These are usually ornamental and add much to the interior design.

STAIR DETAILS

Because of the complexity of stair construction, they are detailed carefully in the

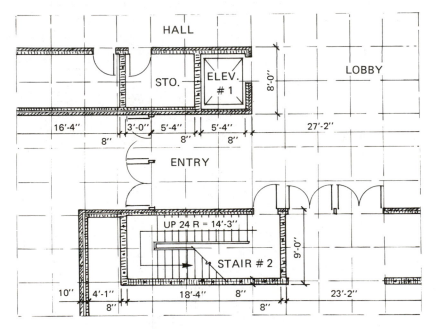

Fig. 29-8 Partial floor plan showing stairs

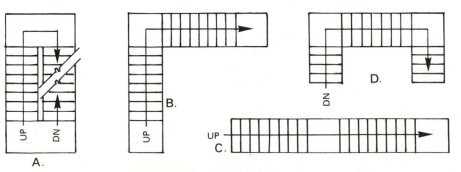

Fig. 29-9 Sample stair layouts

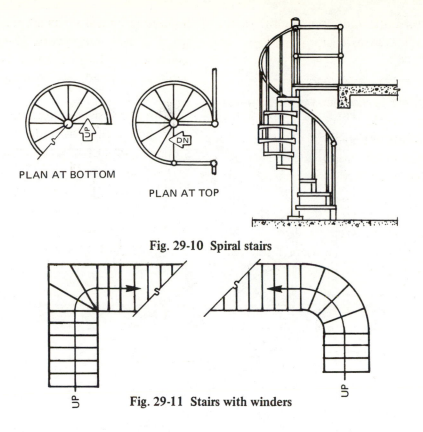

PLAN AT BOTTOM

PLAN AT TOP

Fig. 29-10 Spiral stairs

Fig. 29-11 Stairs with winders

architect's plans. For most commercial buildings, this means special sheets of stair details. The stair section is used to advantage for this purpose, figure 29-12, page 292. The plan of the stair at different levels is usually included on the stair detail sheet. Railing details are shown being an important safety feature as well as a design element.

As noted earlier, when an assembly is fabricated away from the job, shop drawings are prepared. The carpenter may be responsible for preparation and layout of the structure to receive the stair assemblies. The carpenter may also help set the prefabricated unit, and in the case of wood stairs, may make the final assembly on the job. While it is not the purpose of this book to teach stair building, the following gives some basic information about stairs. A few of the important terms and features found on blueprints relating to stairs are given in figure 29-13, page 293.

When a floor plan indicates a stair with 15 R = 9'-2" it means there are 15 equal risers totaling 9'-2". The run of this flight would be 14 times the tread dimension. The number of treads is always one less than the number of risers in a stair assembly. The run of the stair flight may be indicated or the tread dimension given. Rules generally used to proportion the riser and tread for a given flight are as follows:

- Two times the riser plus the tread equals 24" to 25".
- The riser times the tread equals 70" to 75".
- The riser plus the tread equals 17" to 18".

These rules give reasonable proportions. They are based upon the proposition that the same stride length should be maintained whether going up a steep stair or a gentle one. A steep stair with high risers has narrow treads if proportioned by these rules. Interior

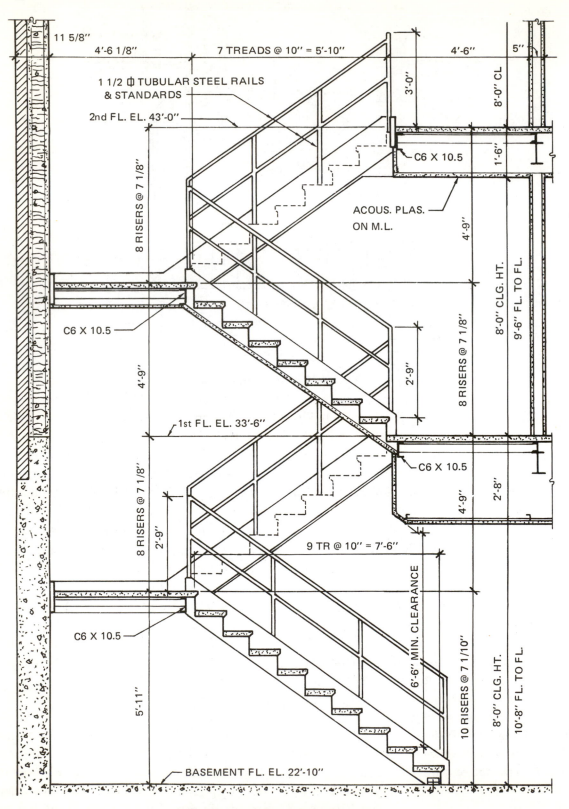

11 5/8"

4'-6 1/8"   7 TREADS @ 10" = 5'-10"   4'-6"   5"

1 1/2 ⌀ TUBULAR STEEL RAILS
& STANDARDS

3'-0"

8'-0" CL.

2nd FL. EL. 43'-0"

C6 X 10.5

1'-6"

8 RISERS @ 7 1/8"

ACOUS. PLAS.
ON M.L.

4'-9"

C6 X 10.5

4'-9"

2'-9"

8 RISERS @ 7 1/8"

8'-0" CLG. HT.

9'-6" FL. TO FL.

1st FL. EL. 33'-6"

C6 X 10.5

8 RISERS @ 7 1/8"

2'-9"

4'-9"

2'-8"

9 TR @ 10" = 7'-6"

C6 X 10.5

6'-6" MIN. CLEARANCE

10 RISERS @ 7 1/10"

5'-11"

8'-0" CLG. HT.

10'-8" FL. TO FL.

BASEMENT FL. EL. 22'-10"

Fig. 29-12 Architectural stair section

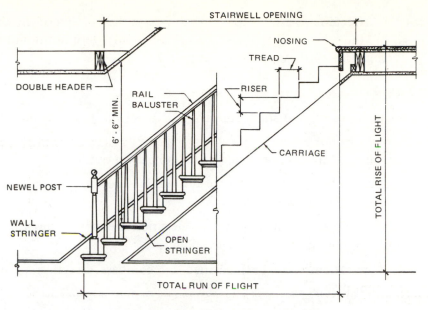

Fig. 29-13 Stair construction terms

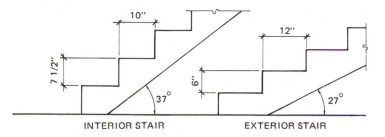

Fig. 29-14 Typical stair step proportions

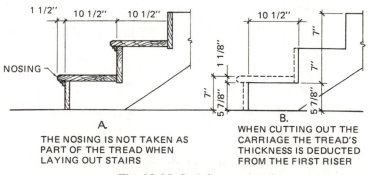

A.
THE NOSING IS NOT TAKEN AS
PART OF THE TREAD WHEN
LAYING OUT STAIRS

B.
WHEN CUTTING OUT THE
CARRIAGE THE TREAD'S
THICKNESS IS DEDUCTED
FROM THE FIRST RISER

Fig. 29-15 Stair layout details

steps have risers of seven to eight inches while exterior entrance step risers are closer to six inches, figure 29-14.

The tread and riser dimensions selected are the basis for the stair layout and establish the slope of the stair. A single line drawing is often used to show this layout. The nosing projects beyond this layout tread dimension but does not enter into the calculations, figure 29-15. The total width of the tread surface is, of course, affected. Also note that the thickness of the tread must be considered when cutting the carriage. The first riser at the bottom of the flight would be too high if this were not done, figure 29-15b, page 293.

SUMMARY

- Cabinets, fixtures, and stairs are rarely built on the job in commercial work.

- Base cabinets show on floor plans as a simple solid outline.

- Wall cabinets show on floor plans as dotted lines because they are above the plane used to draw the plan.

- Typical cross sections and complete elevation views give most of the information required to construct cabinets.

- Cabinet elevations may be identified by compass direction or by a keyed floor plan.

- Field measurements are made to verify building dimensions before final shop drawings are made.

- Standardized cabinets are fabricated in units of varying sizes which are made to fit building dimensions by means of filler strips.

- Cabinet tops for standardized cabinet layouts are the same as for custom cabinets.

- Stair designs and details are carefully worked out by the architect.

- When space is limited, spiral stairs and stairs with winders are used.

- The number of treads in a flight of stairs is always one less than the number of risers.

- Stair treads and risers are proportioned by rules which maintain the same stride length for both steep and gentle flights.

- The nosing on the tread is not included when calculating total run of a flight of stairs.

REVIEW QUESTIONS

1. How are counter-high base cabinets shown on floor plans?

2. How are wall cabinets mounted above the base cabinets shown on floor plans?

3. Name two other views besides the plans that are used to fully describe the cabinets.

4. Complete the elevation sketch of the wall cabinet below by indicating the door swings if they open in the middle. Also show two shelves equally spaced.

5. The main task of the commercial carpenter when finishing the interior of a building is:

   a. to build the cabinets and millwork.

   b. to install shop-built cabinets and millwork.

6. Standardized manufactured cabinets are made in sizes based upon multiples of:

   a. 2 inches.   c. 4 inches.

   b. 3 inches.   d. 6 inches.

7. How are standardized cabinets made to fit when placed between two walls?

8. Rules used to proportion risers and tread dimensions are based upon what proposition?

9. Name two types of stairs that can be used when floor space is limited.

10. Why are special sheets of drawings of stair details usually included in commercial building plans?

11. Interior stairs have riser dimensions of

    a. 5 to 6 inches.   c. 7 to 8 inches.

    b. 6 to 7 inches.   d. 8 to 9 inches.

12. How many treads does a stair flight have if there are 12 risers?

Unit 30
Exterior Finish

OBJECTIVES

After studying this unit, the student should be able to:

- identify window types by name and tell how their sashes operate.
- tell how lightweight metal curtain wall panels are constructed and installed.
- give two types of cornice work as related to the architectural styling of the building.

DOORS AND WINDOWS

Exterior doors are listed in the door schedules with all the other doors. They require more special hardware than interior doors. The door closer is a good example. Fancy entrances sometimes use floor mounted closers, figure 30-1. A blockout must be made in the floor for these closers. Accurate layout and forming during earlier floor construction stages are important for efficient setting of these doors. Special detail drawings of entrances usually indicate any unusual preparation required. The specifications are also an important source of information. More detailed treatment of doors and frames was included in Unit 28.

Windows are made in many types. Working drawings, especially the elevation views, represent them carefully. The architect wants them to look like the real thing on the drawings because the shape and pattern of the windows are important design elements.

To further identify them, some window types have graphic conventions added. The most important one is the addition of dotted lines to indicate the hinged side of a swinging sash, figure 30-2. The same dotted line indication is used on hopper vents, hinge side at the lower edge and awning sash hinged at the upper edge, figure 30-3. The two dotted lines always meet at the center of the hinged side of the sash. Horizontal sliding sash and sliding glass doors may have arrows added to indicate the movable sash, figure 30-4.

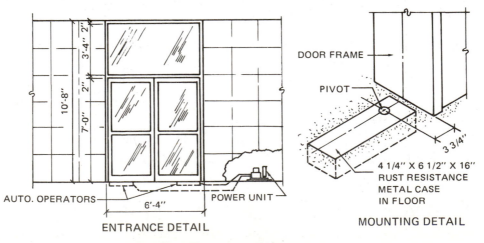

ENTRANCE DETAIL

MOUNTING DETAIL

Fig. 30-1 Floor-mounted door closer

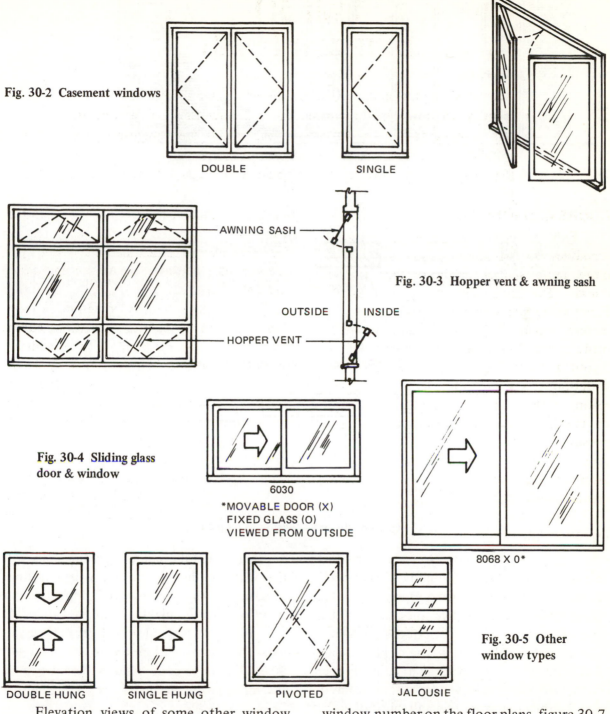

Fig. 30-2 Casement windows

DOUBLE   SINGLE

AWNING SASH

Fig. 30-3 Hopper vent & awning sash

OUTSIDE   INSIDE

HOPPER VENT

Fig. 30-4 Sliding glass door & window

6030
*MOVABLE DOOR (X)
FIXED GLASS (O)
VIEWED FROM OUTSIDE

8068 X 0*

DOUBLE HUNG   SINGLE HUNG   PIVOTED   JALOUSIE

Fig. 30-5 Other window types

Elevation views of some other window types are shown in figure 30-5. The windows are sometimes identified on the elevation views by means of a window schedule symbol, figure 30-6, page 298. They are always identified with a symbol or else with the manufacturer's window number on the floor plans, figure 30-7, page 298.

The plan view of any building is a horizontal section taken at a plane passing through the windows. The horizontal section through the window, being small scale, is more symbolic

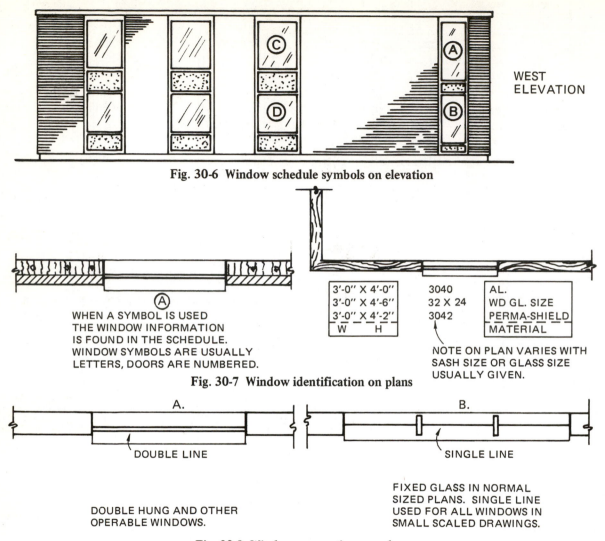

Fig. 30-6 Window schedule symbols on elevation

WEST ELEVATION

Ⓐ WHEN A SYMBOL IS USED THE WINDOW INFORMATION IS FOUND IN THE SCHEDULE. WINDOW SYMBOLS ARE USUALLY LETTERS, DOORS ARE NUMBERED.

3'-0" X 4'-0"	3040	AL.
3'-0" X 4'-6"	32 X 24	WD GL. SIZE
3'-0" X 4'-2"	3042	PERMA-SHIELD
W H		MATERIAL

NOTE ON PLAN VARIES WITH SASH SIZE OR GLASS SIZE USUALLY GIVEN.

Fig. 30-7 Window identification on plans

A.

DOUBLE LINE

B.

SINGLE LINE

DOUBLE HUNG AND OTHER OPERABLE WINDOWS.

FIXED GLASS IN NORMAL SIZED PLANS. SINGLE LINE USED FOR ALL WINDOWS IN SMALL SCALED DRAWINGS.

Fig. 30-8 Window conventions on plans

than might be expected. No particular effort is made to show all the window parts as cut. Usually a window that opens, especially double hung, is represented by two lines, figure 30-8a. They are placed in the depth of the wall about where they actually occur. The sill line outside the window is usually drawn in its proper position. The inside window stool is rarely shown projecting from the wall, but rather the wall line is continued unbroken, figure 30-8.

Another common type window is the fixed glass panel. This type is indicated by a single line across the window opening in the plan view, figure 30-8b. In all cases, good

drafting technique emphasizes the walls with the window details shown by lighter lines. This makes the window and door openings stand out clearly, making them easy to locate.

Window schedules give the necessary data for identification and ordering, figure 30-9. The large scale window details provide the information needed to set the windows, figure 30-10.

Shop drawings are not prepared for standard windows since manufacturers furnish basic sections in catalogs. These catalog drawings are often traced by the architects in preparing the detail sections.

MARK	NO. REQ'D	SIZE OR MFG NO.	TYPE	RFG. OPNG	REMARKS
Ⓐ	6	32 x 24 2LT GL. SIZE	D.H. WD.	3'-4'' x 4'-10''	OBSCURE GL.
Ⓑ	2	3040	S.H. AL.	3'-0 1/2'' x 4'-0 1/2''	
Ⓒ	2	4020	H.S. AL.	4'-0 1/2'' x 2'-0 1/2''	
Ⓓ	4	3N5	CAS. WD.	5'-1'' x 5'-5 7/8''	ANDERSEN
Ⓔ	8	2820	BSMT. WD.	2'-8 5/8'' x 1'-11 1/4''	
Ⓕ	2	26'' x 80'' GL. SIZE	FIXED	2'-4'' x 6'-10''	DOUBLE GL.

Fig. 30-9 Window schedule on plans

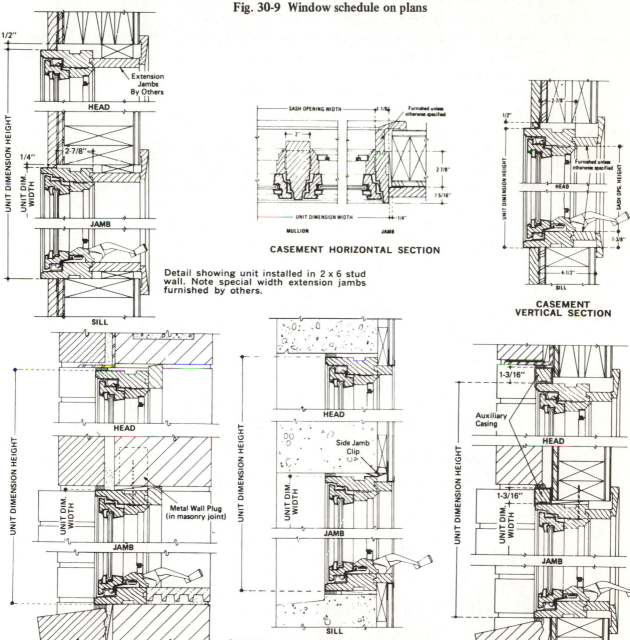

Detail showing unit installed in 2 x 6 stud wall. Note special width extension jambs furnished by others.

CASEMENT HORIZONTAL SECTION

CASEMENT VERTICAL SECTION

Casement unit installed in a 12" masonry wall. Unit secured through side jambs into metal wall plugs located in masonry joints.

Casement unit installed in a pre-cast masonry wall with interior wall on furring strips. Metal trim and extension jambs used on interior. Unit secured in opening with side jamb clips nailed to furring strips.

Detail showing Auxiliary Casing applied for a wider casing affect or for remodeling to arrive at opening widths.

Fig. 30-10 Typical installation details for casement windows

CURTAIN WALLS

Special windows and curtain walls require shop drawings before they can be made. *Curtain walls* are non-load-bearing exterior walls. They may be made of almost any material. Masonry curtain walls are mentioned in Unit 23. Lightweight metal framed curtain walls are made in sections, some with windows. The solid portions are insulated, and in cold climates the glass may be doubled. Many different finishes are used in addition to the glass sections. Porcelain steel panels, plastic faced plywood, cast stone, natural exposed-aggregate panels, and many more are used. Curtain walls conserve floor space and give the building a neat and ordered appearance, figure 30-11. Panels are often fabricated two stories high. They are attached to the building by special clips that are secured by bolting or welding, figure 30-12.

SIDING

Low-rise buildings in suburban zones may have conventional siding detailed. The elevation drawing represents the siding realistically. A note giving the type of siding is usually included. Several types of siding indications are shown in figure 30-13. The exact material required is found in the specs which also give instructions for its application.

ROOFING

Carpenters apply shingle-type roofing materials and some roll roofing on low-slope roofs. Built-up roofing, sometimes called tar and gravel roofing, is used on flat roofs. This type roofing, as well as sheet-metal roofing, is done by roofers, never by carpenters.

Shingle type roofs have slopes of three inches or more per foot of run. The roof surface shows in the elevation views. The

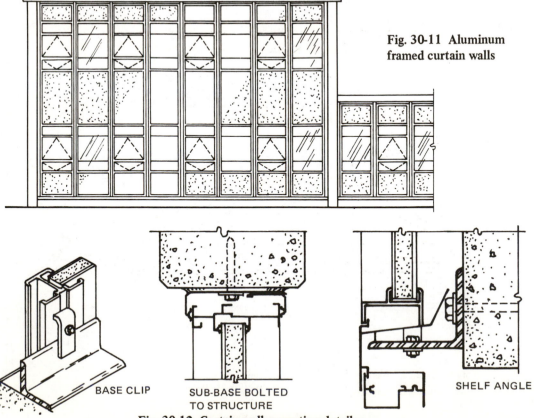

Fig. 30-11 Aluminum framed curtain walls

BASE CLIP

SUB-BASE BOLTED TO STRUCTURE

SHELF ANGLE

Fig. 30-12 Curtain wall mounting details

shingle coursing is indicated by closely spaced parallel lines, figure 30-14. Usually only a part of the roof surface is so treated, and a note identifies the type material. Larger scaled sections show the roofing in more detail together with the material and structure supporting it, figure 30-15. The specifications must be checked for exact material and installation information.

CORNICE WORK

The exterior ornamental trim on a building at the juncture of the wall and the roof is called the *cornice*. Wide projecting cornices

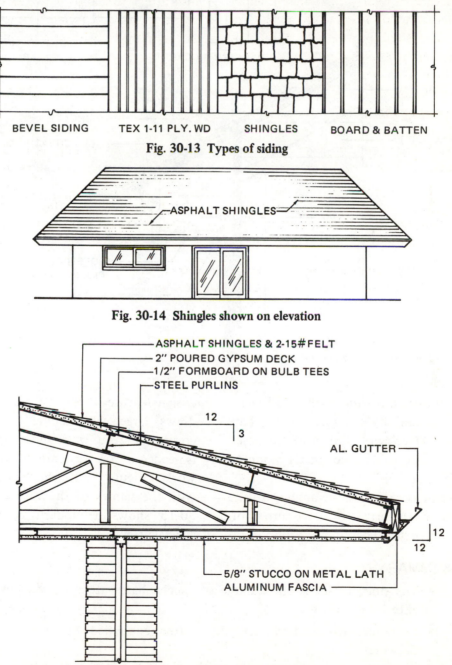

BEVEL SIDING   TEX 1-11 PLY. WD   SHINGLES   BOARD & BATTEN

Fig. 30-13 Types of siding

ASPHALT SHINGLES

Fig. 30-14 Shingles shown on elevation

ASPHALT SHINGLES & 2-15#FELT
2" POURED GYPSUM DECK
1/2" FORMBOARD ON BULB TEES
STEEL PURLINS

12
3

AL. GUTTER

12
12

5/8" STUCCO ON METAL LATH
ALUMINUM FASCIA

Fig. 30-15 Roofing & eave detail

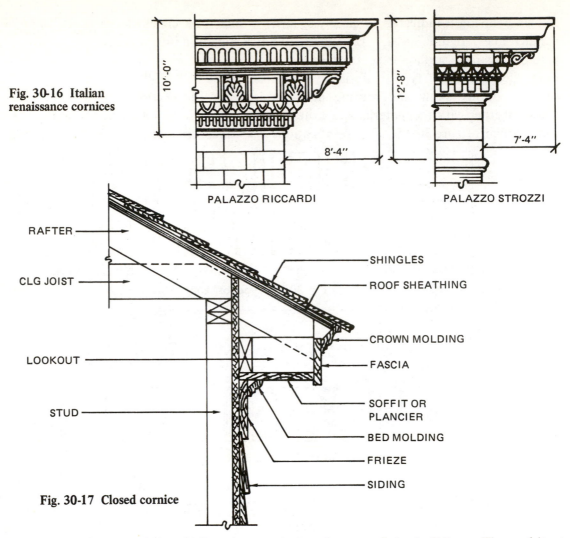

Fig. 30-16 Italian renaissance cornices

PALAZZO RICCARDI

PALAZZO STROZZI

Fig. 30-17 Closed cornice

made of stone are a feature of the old Renaissance architectural style, figure 30-16. Late 19th century American buildings often copied the stone cornice using sheet metal mounted on wood framing.

The term cornice is also applied to the finish work at the eaves of a modern roof. This eave or cornice detail is an important design feature of the building. The architect draws large-scale detailed sections of the cornice. Careful work and attention to details is needed by the carpenter in doing this cornice work. Figure 30-17 shows a cornice section with the names of the various parts. Figure 30-15 shows a modern eave designed for special effects.

SUMMARY

- The door closer is an example of special hardware usually found on exterior doors only.
- Entrances to buildings are most often explained by special detail drawings.
- Windows are drawn carefully on the elevations, and some have graphic symbols besides to help identify them.

- Operable windows are usually shown by a double-line symbol on floor plans.

- Fixed glass is shown as a single line on floor plans.

- Shop drawings are not required for standard windows.

- Metal-framed curtain walls conserve floor space and give the building a neat and ordered appearance.

- Windows and fixed glass panels are often included in curtain-wall panels.

- The cornice is an important design feature of a building.

- Careful work is required when the carpenter builds cornices, and the detail drawings must be carefully followed.

REVIEW QUESTIONS

1. What item of hardware is often required on exterior doors but rarely on interior ones?

2. Identify these elevation views of common window types.

   a.
   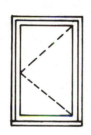

   c.
   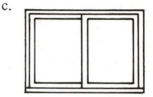

   b.
   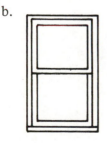

   d.
   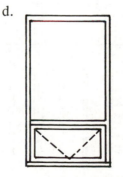

3. How are windows identified on floor plans?

4. Complete the following window drawings as found on floor plans.

   a.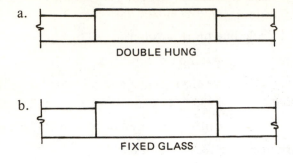

   DOUBLE HUNG

   b.

   FIXED GLASS

5. What are curtain walls?

6. Name three advantages or good features of curtain walls constructed in panels with metal frames.

7. Name the two types of wood siding indicated in the sketches below.

   a.

   b.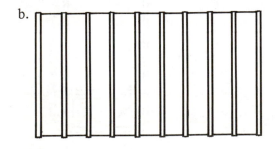

8. What type of roofing is never applied by the carpenter?

9. What is a minimum practical slope for shingle type roofs?
a. 2 inches per foot.	c. 4 inches per foot.
b. 3 inches per foot.	d. 6 inches per foot.

10. Match these cornice part names with their proper parts in the sketch below (frieze, fascia, soffit, lookout, quarter-round).

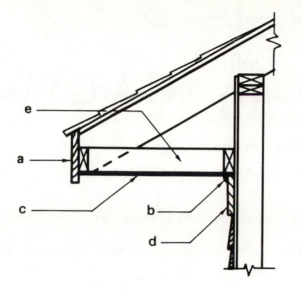

MECHANICAL AND ELECTRICAL WORK

UNIT 31
PLUMBING, HEATING, AND A/C WORK

OBJECTIVES

After studying this unit, the student should be able to:

- identify symbols for plumbing fixtures and piping.

- give the reason for having traps and vents in a plumbing system.

- explain why tall buildings are divided into zones for mechanical services.

PLUMBING SYSTEMS

Systems installed by plumbers include water service and distribution, sanitary drainage, storm drainage, and gas piping. It is not uncommon to find all these systems shown on the plumbing plans of a building. Fire protection systems include standpipes, hose cabinets, and automatic sprinkler systems. These sprinkler systems are usually installed by special mechanical subcontractors and the drawings are separate from the normal plumbing plans.

WATER SERVICE AND DISTRIBUTION SYSTEMS

Pure water, suitable for drinking, is piped into every modern building through a service line from the street main. The size of the line is determined after making an analysis of the fixture demands in the building. Most water piping terminates at plumbing fixtures placed at convenient locations throughout the building. Plumbing fixtures are easy to identify on drawings since their symbols, as used on floor plans, look much like top views of the fixtures, figure 31-1. Architectural floor plans show plumbing fixtures wherever they occur throughout the building. Plumbing plans show the fixtures together with their connecting systems of piping.

Two piping distribution systems are used for clean water. One is for cold water and one is for hot water. Since many fixtures have both hot and cold water outlets, it is necessary to clearly identify the piping systems. This is done by using standard symbols for each system as shown in figure 31-2.

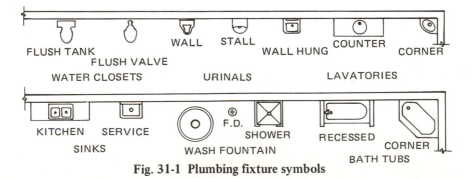

Fig. 31-1 Plumbing fixture symbols

COLD WATER

HOT WATER

HOT WATER RETURN

GATE VALVE W/ UNION

GLOBE VALVE W/ELBOW
(TURNED DOWN)

CHECK VALVE W/ TEE

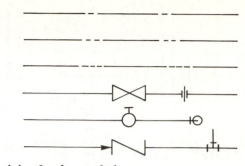

Fig. 31-2 Water piping & valve symbols

The main service line usually enters the building at the mechanical room where it is divided into several branches. In large buildings, the fire protection system has priority and water is often stored in tanks for emergencies. Another branch supplies the cold water to the fixtures. A third branch connects to the water heater which supplies the domestic hot water for the building.

To make the hot water systems work well, that is, to have hot water at each fixture all the time, the hot water is usually circulated. A small pump, or in some cases natural convection flow, causes the hot water to circulate all the time. A return pipe is required in order to make this circulating loop complete. This return pipe has a special symbol shown in figure 31-2. Common valves used in water distribution systems are also included in figure 31-2.

SANITARY DRAINAGE SYSTEMS

All plumbing fixtures supplied with water must have some system to carry off the excess water and waste that accumulate. Floor drains also connect to this system which is called the *sanitary drainage system.* The term sanitary is misleading because the water conveyed by this system is contaminated. The degree of contamination is reflected in two classifications of piping. The discharge from fixtures such as water closets and urinals is conveyed in soil pipes and soil stacks. The piping serving other fixtures is called waste piping. No distinction is made in the symbol used for drainage piping whether soil or waste, figure 31-3. Sometimes they are labeled to indicate the difference.

Every fixture that connects to the sanitary drainage system must have a trap. A *trap* is a fitting or device that provides a liquid seal to prevent sewer gasses from discharging into the room, figure 31-4, page 308. In order for the traps to work properly, the systems must be vented. Vent piping normally conveys only air and gasses and extends to the open air above the roof. On plumbing plans the vent piping is indicated by dotted

SOIL OR WASTE PIPE

VENT PIPE

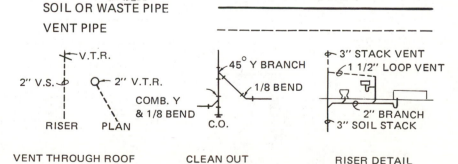

VENT THROUGH ROOF   CLEAN OUT   RISER DETAIL

Fig. 31-3 Drainage piping & fitting symbols

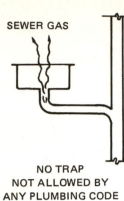
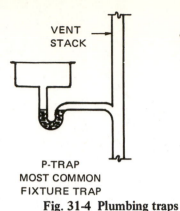

Fig. 31-4 Plumbing traps

lines, figure 31-3, page 307. Other drainage symbols and notations are given in figure 31-3 as well.

STORM DRAINAGE SYSTEM

A *storm drainage system* is one that conveys rain water, surface water, cooling water, or similar liquid wastes. Since storm water is clean when compared to sewage, it does not require much treatment. If storm water is allowed to enter the sewer, it greatly increases the load on the sewer plants. (Most city codes do not allow storm water to be drained into the sanitary sewer.) Separate storm sewers and storm drainage systems are required when possible.

Roof drains are used in the construction of most commercial buildings, especially the large flat-roofed buildings. Footing and subsurface drains control ground water. Area drains, yard drains, etc. all connect to the storm sewer in the street. No traps are required in storm drainage systems and consequently no vents are used. Piping symbols are generally the same as those used for sanitary drainage with the distinction being made by notes, figure 31-5.

Important to the proper operation of any drainage system is the slope of all parts of the system. (See Unit 10 for methods of calculating slope.) The drainage piping has first priority for space in the building and where there is a problem, other systems must detour. The location of holes for drainage piping must be carefully checked when forming for concrete walls and floors.

GAS PIPING

Gas piping is limited in many buildings to the service pipe, gas regulator, and simple distribution system in the mechanical room. Gas-fired equipment for hot-water or steam

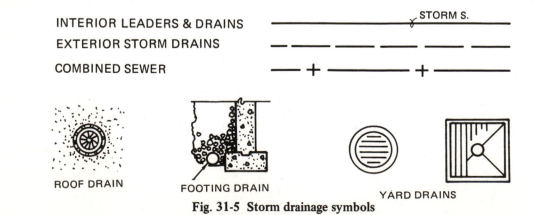

Fig. 31-5 Storm drainage symbols

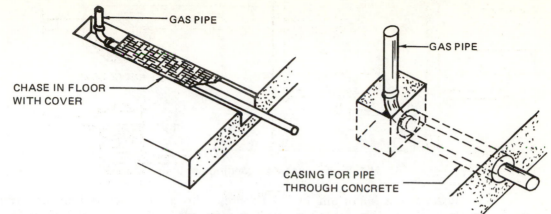

Fig. 31-6 Gas piping in concrete floor

heating systems is located there. When gas-fired space heaters are distributed around the building, a more extensive system is required.

Of special interest are the precautions taken when installing the gas service line into the building. The pipe should rise out of the ground before entering the wall in order to vent any gas leaking out of the pipe underground. Also, a gas line cannot be installed under or in a concrete slab on grade. If it is necessary to locate the gas line in the slab, a casing or sleeve large enough to allow replacement of the gas line is used. A channel or chase formed in the floor with a suitable cover is another possibility, figure 31-6.

Inside the building the gas lines are run exposed, in hollow partitions, or concealed in chases. The piping is wrought iron or steel pipe, often called black iron pipe. Water piping is usually galvanized steel or copper so the gas piping is easy to recognize. The gas piping symbol used on plans is a solid line broken at intervals with the letter G inserted. The size is noted along side of the pipe as is common in all piping drawings, figure 31-7.

HEATING AND A/C WORK

Year-round air conditioning, climate control, and space conditioning are other names for the process of treating air to control comfort in a space. This includes (simultaneously) air temperature, humidity, cleanliness, and distribution. Although a number of systems are used which heat only in winter, the comfort demands of large commercial buildings require more elaborate systems.

Industrial and low commercial buildings can be conditioned by rooftop heating and cooling units. A minimum amount of ductwork is required since several units can be located over the roof area, figure 31-8, page 310.

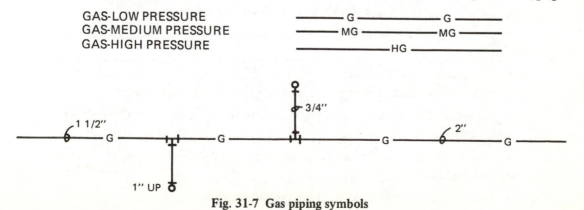

Fig. 31-7 Gas piping symbols

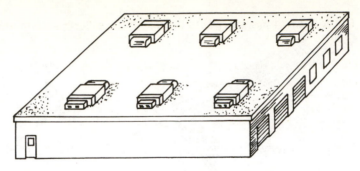

Fig. 31-8 Roof mounted air conditioning units

Dual-duct air systems are important in large buildings including high-rise buildings. Because of the space required for long runs of ducts, tall buildings are divided into zones with separate mechanical rooms for each zone. High velocity systems have been developed to reduce the duct sizes also.

Duct layouts on mechanical plans are quite detailed. Standard symbols and conventions are shown in figure 31-9. Ducts that run horizontally are hung from the structural slab and concealed with a suspended ceiling. Air ducts, plumbing lines, electrical conduit and fixtures, plus a sprinkler system for fire protection are often hung together in this space.

Fan coil units located around the perimeter of a building and as otherwise required are also used. These units are supplied with hot and chilled water lines which are easier to run than air ducts. Drains for condensate are needed however. The all-electric air conditioning system using through-the-wall heat pumps is popular for hotels, apartments, and some offices. Mechanical plans are hard to read, but the basics given in this unit provide some starting guidelines.

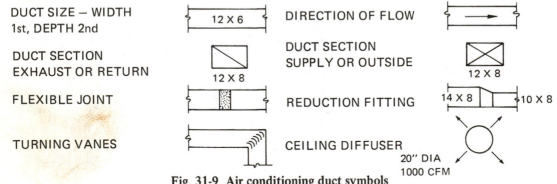

Fig. 31-9 Air conditioning duct symbols

SUMMARY

- Plumbing work includes installation of systems for water service and distribution, sanitary drainage, storm drainage, and gas piping.

- Automatic sprinkler systems are usually installed by special mechanical subcontractors.

- Plumbing plans show the location of all fixtures and the piping systems serving them.

- Symbols are used to identify hot and cold water lines.

- The main water service line usually enters the building at the mechanical room.

- Hot water is circulated through the building by means of a pump and a return pipe forming a loop.

- Sanitary drainage systems convey water and waste from plumbing fixtures to the sewer.

- Traps are used to prevent sewer gasses from escaping into the occupied spaces.

- Vents are needed to ventilate the drainage system and make the traps operate properly.

- A storm drainage system conveys rain water, surface water, cooling water, or similar wastes to the storm sewer.

- Gas piping may not be run in a concrete slab unless placed in a casing or chase.

- The process of controlling air temperature, humidity, cleanliness, and distribution is called year-round air conditioning.

- Tall buildings are divided into zones for mechanical services to keep pipe and air ducts as small and short as possible.

- Mechanical and electrical systems are often installed together in the space above the suspended ceilings.

REVIEW QUESTIONS

1. Hot and cold water distribution systems are found in most buildings. Sketch the piping symbol used for each.

2. How are hot water systems designed to maintain hot water at each fixture at all times?

3. What is domestic hot water?

4. What water system has priority in a large building?

5. What fire protection system is usually installed by a special mechanical subcontractor?

6. Which drainage system conveys relatively clean water, the sanitary drainage system or the storm drainage system?

7. Sketch a P trap connecting a sink to a plumbing stack.

8. Identify the following valves used in plumbing systems.

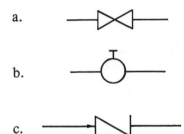

   a.

   b.

   c.

9. Why must plumbing fixtures have traps?

10. What must be done to make the traps work properly?

11. How must a drainage system be installed to operate properly?

12. Sketch the symbol used for gas lines in a building.

13. How can gas lines be safely installed in concrete floors?

14. Name four things included in the process of treating air to control comfort in a space.

15. Why are tall buildings divided into zones for mechanical systems?

16. How are air ducts usually concealed in buildings?

17. Identify these air distribution symbols.

a.

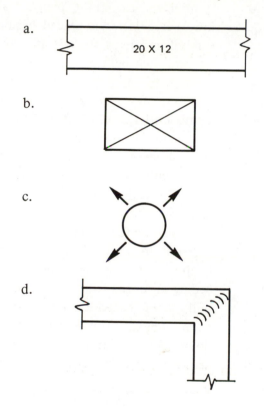

b.

c.

d.

Unit 32
Electrical Work

OBJECTIVES

After studying this unit, the student should be able to:

- outline the electrical system of a building from the power company lines to the various outlets.
- identify lighting outlet symbols used for incandescent and fluorescent fixtures.
- explain how branch circuit wiring is indicated on plans.

ELECTRICAL SYSTEMS

Electrical work in buildings consists of many types. Installing small appliance receptacles and light fixtures is classed as finish work. These are the parts of the system recognized by anyone using the building. Supplying the fixtures is the branch circuit wiring system which is the concealed or rough work. Special power circuits provide for large equipment such as elevators, air conditioning equipment, water heaters, etc. Service entrance equipment feeds the building through switchboards and panels. Besides all these power systems, there are signal systems of many kinds. Learning the symbols used in electrical plans helps one understand this phase of construction.

LIGHT FIXTURES AND RECEPTACLES

Light fixtures are listed in a schedule with each fixture being identified by a capital letter. This letter is noted in the outlet symbol on the floor plans. Sometimes a lower case letter is placed beside the outlet symbol and used to indicate the switch that operates that light.

There are three basic symbols that represent light fixture outlets. They are the ceiling outlet, wall bracket outlet, and fluorescent lamp fixture outlet, figure 32-1. These symbols are used with minor variations for most building light outlets, figure 32-2.

The ceiling outlet symbol can be used for any incandescent lamp holder or fixture. The symbol for a wall mounted bracket fixture has universal application also. Bare-lamp lampholders can be drop cords, porcelain ceiling mounted lampholders, with or without pull chain switches, or swivel base flood lampholders. Any hanging fixture, recessed downlight, or surface mounted enclosed incandescent fixture can also be represented by

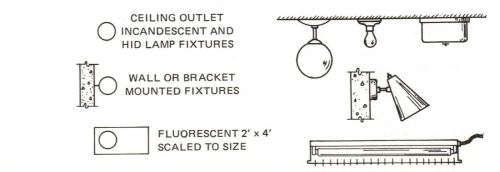

CEILING OUTLET
INCANDESCENT AND
HID LAMP FIXTURES

WALL OR BRACKET
MOUNTED FIXTURES

FLUORESCENT 2' x 4'
SCALED TO SIZE

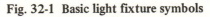

Fig. 32-1 Basic light fixture symbols

the one ceiling outlet symbol. The new high intensity discharge (HID) lamps are also represented by the same symbol. HID lamps include mercury vapor, metal halide, and sodium types.

Fluorescent lamp fixtures are much larger than incandescent fixtures and are drawn to scale. One outlet symbol is used for a single fixture or a continuous row of similar fixtures, figure 32-2. Fluorescent fixtures may be bare lamp strip lights or open reflector types. Better fixtures have two-way shielding of the lamps. Suspended ceiling installations usually have recessed fluorescent fixtures designed to fit the ceiling grid.

How the lighting is controlled is the next important information the blueprint reader needs to determine. The wall switch symbol S is used to show the location in the room of local control switches. As noted above, a lowercase letter can be used to indicate the light fixture being controlled. This is not needed if the wiring connections are simple and obvious. Several variations of the basic switch symbol are shown in figure 32-3. Also included are symbols for remote switches and dimmers.

Receptacles or convenience outlets are wiring devices that permit the connection of attachment plugs. Movable lamps and small appliances are connected to these receptacles which are located at convenient spots in the rooms. Heavy duty receptacles are available for special equipment such as electric dryers. Receptacle symbols are based upon variations of the basic outlet symbol, the small circle. Several are shown in figure 32-4, page 316. An electrical *outlet* is a point in the system that provides for the connection of a power consuming or control device. It is where the concealed wiring is brought to the surface to be used.

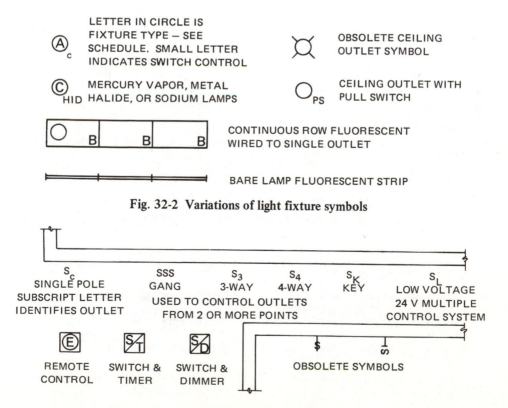

Fig. 32-2 Variations of light fixture symbols

Fig. 32-3 Switch symbols

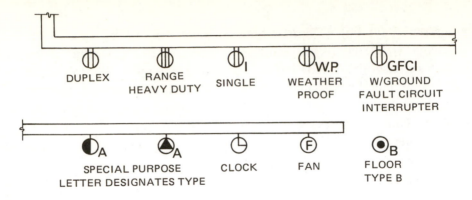

Fig. 32-4 Receptacle symbols

BRANCH CIRCUIT WIRING

The outlets, which are seen in the building, connect to the power through branch circuits. Because of the hazard involved in the use of electricity, the loads are divided up into small units called branch circuits. These small circuits require modest-sized wires and are protected by fuses or circuit breakers against overloads. More protection to the user is provided by a system of grounding. For the best protection, some receptacles are now equipped with ground fault circuit interrupters (GFCI). Shock from defective equipment is avoided by this means.

The branch circuits are shown on electrical plans with symbols as in figure 32-5. These symbols usually represent conduit with electrical wires pulled through them. The number of wires is indicated by the slash marks on the conduit symbol.

The various outlets served by a circuit are interconnected and the circuit is numbered.

The circuit is not connected back to the panelboard but instead is aimed in that direction and suitably noted. To conserve conduit and wire, this *home run* (wiring between first outlet and the panel) may have more than one circuit, figure 32-6.

Special branch circuits are usually run to each piece of electrical equipment and are classed as power or heavy duty circuits. Electrical equipment symbols are given in figure 32-7.

SERVICE ENTRANCE AND DISTRIBUTION EQUIPMENT

The service lines from the utility company power lines may enter the building overhead, but for most modern commercial buildings, the lines enter underground. *Primary Service* is high-voltage service, 2400 V and up. It enters the building in rigid steel conduit ending in a service transformer. A *transformer* is an electrical device that changes electricity

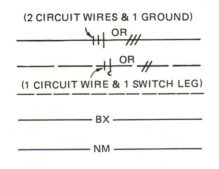

CONDUIT CONCEALED IN CEILING OR WALLS

CONDUIT CONCEALED IN FLOOR

CONDUIT EXPOSED

BX - ARMORED CABLE

NON-METALIC CABLE (ROMEX)

Fig. 32-5 Branch circuit wiring symbols

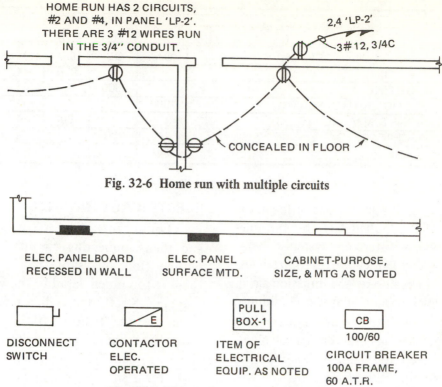

HOME RUN HAS 2 CIRCUITS, #2 AND #4, IN PANEL 'LP-2'. THERE ARE 3 #12 WIRES RUN IN THE 3/4" CONDUIT.

2,4 'LP-2'

3# 12, 3/4C

CONCEALED IN FLOOR

Fig. 32-6 Home run with multiple circuits

ELEC. PANELBOARD RECESSED IN WALL

ELEC. PANEL SURFACE MTD.

CABINET-PURPOSE, SIZE, & MTG AS NOTED

DISCONNECT SWITCH

CONTACTOR ELEC. OPERATED

PULL BOX-1
ITEM OF ELECTRICAL EQUIP. AS NOTED

CB
100/60
CIRCUIT BREAKER 100A FRAME, 60 A.T.R.

Fig. 32-7 Electrical equipment symbols

from one voltage to a different voltage. The service transformer would be a step-down transformer reducing the high voltage (2400V and up) to a lower voltage (480V to 120V). This lower voltage is called the *secondary service* and would be the main service in buildings without transformers.

Figure 32-8 shows the arrangement of the components of the electrical system of a typical commercial building. Feeders from the main switchboard supply panels located at convenient places in the building. The

smallest branch circuits are fed from lighting and small appliance panels. These panels have gangs of circuit breakers or fused switches which are numbered and correspond to the circuit number on the floor plans. Electrical plans have schedules showing the details of each panel, figure 32-9, page 318.

SIGNAL EQUIPMENT

A number of other electrical devices are used in buildings that require wiring systems. The telephone is a good example. Most plans

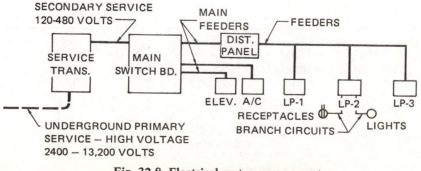

SECONDARY SERVICE 120-480 VOLTS

MAIN FEEDERS

FEEDERS

DIST. PANEL

SERVICE TRANS.

MAIN SWITCH BD.

ELEV. A/C   LP-1   LP-2   LP-3

RECEPTACLES
BRANCH CIRCUITS

LIGHTS

UNDERGROUND PRIMARY SERVICE – HIGH VOLTAGE 2400 – 13,200 VOLTS

Fig. 32-8 Electrical system components

LIGHTING PANEL LP-2		120/208V. 3φ, 4W 60 HZ, 200 AMP				
NO	SERVES	φA	φB	φC	POLES	TRIP
1	LIGHTING			900	1	20A
2	LIGHTING		1600		1	
3	FLOOD LIGHTS - SOUTH	1500			1	
4	SIGNS & M.V. LIGHTS			1500	1	
5	RECEPTACLES		900		1	
6	SPARE	850			1	

Fig. 32-9 Panel schedule

show empty conduit runs to phone locations because the wire is pulled later by the telephone company. Intercom systems, bells, buzzers, etc. are other types of communication systems. Fire, smoke, and intrusion alarm systems are now common in modern buildings.

The plans for these signal systems are separate from the general electrical plans. Power is furnished for them from circuits so designated. Symbols used on plans for this equipment are given in figure 32-10.

EMERGENCY EQUIPMENT

Exit lights are required in all commercial buildings. The power for these lights is taken off the secondary feeder ahead of the main switches. This is done to provide emergency lights and exit signs in case all electricity is turned off in the building. In case of complete power failure, hospitals require auxilliary power services such as batteries for exit lights and generators for critical services (see Appendix).

ELECTRICAL DRAWINGS

Equipment, light fixtures, receptacles, and their connecting circuits are shown on outline floor plans of the building. When there is too much detail to show all clearly on one plan, a reflected ceiling plan can be used to locate the lighting fixtures. The other outlets and equipment are shown on a regular electrical plan. The panel details are shown in chart form and are often placed on a separate sheet. The overall distribution system is outlined in a riser diagram using single-line or block diagram drafting techniques. Finally, the signal equipment plans for each floor, plus details, are drawn. All these sheets carry the same prefix in their sheet numbers, the letter E, (E01, E02, E15).

All electrical drawings for buildings are greatly dependent upon symbols. Many of the common symbols used are illustrated in this unit. A more complete list is printed in the Appendix. There is no universal standard for these symbols and they may, and often

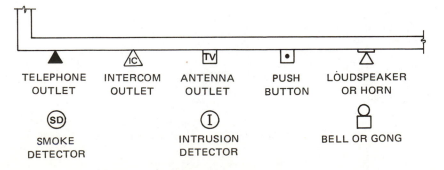

Fig. 32-10 Signal system symbols

do, change from job to job. Included in each set of plans is a list of the symbols used. A thorough study of these symbols is a must for anyone reading electrical plans (see Appendix).

SUMMARY

- Electrical work in buildings consists of rough or concealed work and finish work.

- Service entrance equipment feeds the building through switchboards and panels.

- Symbols are used extensively in electrical plans.

- Light fixtures are identified by means of capital letters keyed to a fixture schedule.

- Ceiling outlet symbols are used for all types of incandescent and HID light fixtures.

- Fluorescent lamp fixtures are drawn to scale on electrical plans.

- The letter S is the symbol used to represent wall switches.

- Receptacles are wiring devices that permit the connection of attachment plugs.

- Branch circuits feed electricity to lights, receptacles, and other outlets.

- Circuit breakers, systems of grounding, and ground fault circuit interrupters make the use of electricity safer.

- Primary service is high-voltage service and must be reduced by a transformer.

- Secondary service is low-voltage suitable for general distribution throughout a building.

- Signal systems include telephones, intercoms, bells, fire, smoke, and instrusion alarms.

REVIEW QUESTIONS

1. What work classed as finish electrical work is done by the electrician?

2. Why are electrical systems divided up into small units called branch circuits?

3. Identify these light fixture symbols:

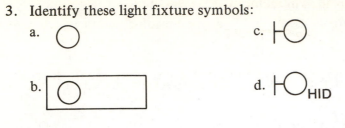

a.

b.

c.

d. HID

4. What does the capital letter placed on a fixture symbol mean?

5. Draw the following switch symbols:

    a. single pole wall switch

    b. 3-way wall switch

6. What is a 3-way switch used for?

7. Identify the following receptacles:

a.

b.

c.

d.

8. How are branch circuits protected from overloads?

9. What does this wiring symbol mean?

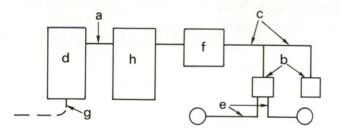

2,4

LP-2

10. How is the number of wires indicated on conduit runs? Illustrate.

11. Match the following names to the electrical system parts shown below.

   primary service   distribution panel
   secondary service   branch circuits
   transformers   lighting panels
   switchboard   feeders

12. Sketch these symbols as used on signal system plans.
   a. telephone outlets

   b. smoke detector

   c. TV antenna outlet

13. When would light fixtures be shown on a reflected ceiling plan instead of on the floor plan?

14. Why is it necessary to study the symbol list given in each set of electrical plans?

Glossary

The following are selected words and terms used in *BLUEPRINT READING FOR COMMER-CIAL CONSTRUCTION.* For a comprehensive list of construction terms see *CONSTRUCTION DICTIONARY,* National Association of Women in Construction, P. O. Box, Phoenix, Arizona 85005. For architectural abbreviations see *DICTIONARY OF ARCHITECTURAL ABBREVIATIONS SIGNS AND SYMBOLS.* New York: The Odyssey Press, 1965.

A/C: Alternating electrical current.

AGC: The Associated General Contractors of America.

AIA: The American Institute of Architects.

Airspace: A small cavity separating parts of a wall or other building assembly, often 1″ to 2″.

AISC: American Institute of Steel Construction.

Angle iron: A hot-rolled structural steel shape with two equal or unequal legs joined at right angles, symbol ∟.

Architect: A professionally qualified and duly licensed designer and administrator of building construction projects.

ASTM: American Society for Testing and Materials.

Auxiliary view: A special view projected from the object to give the true size and shape of a sloping surface.

Awning window: An outswinging window consisting of one or more sashes hinged at their top edges.

Balloon framing: A system of framing wooden buildings using closely spaced 2 x 4 studs extending in one piece from the foundation sill to the highest plate. Intermediate floor joists are supported by ribbon boards recessed into the edge of the studs.

Bars: Hot-rolled structural steel sections eight inches or less in width of rectangular, square, round, or other regular shape.

Base cabinet: A kitchen cabinet 34 to 36 inches high, 24 inches deep, with a work surface or top. Sinks and cooking units are installed in base cabinets.

Batter boards: Temporary horizontal supports erected near the corners of a building layout to hold lines representing the building outline.

Beam: A horizontal structural member spaced 48 inches or more apart and carrying loads principally in bending. Beams fall between joists and girders in their relative importance in a structure.

Bench mark: A fixed point of known or assumed elevation used to establish a reference plane for a project.

Billet: A steel slab used as a base plate to distribute column loads to the foundation.

Blueprint: A copy in negative form of original drawings with white lines on a blue background. Commonly used to mean prints made by any process including Diazo.

Bond beam: A horizontal reinforced concrete or concrete masonry member built into or at the top of a masonry wall to strengthen it. U-shaped masonry units used eliminate the need for forms.

Bonding: The technique of tying masonry units and assemblies together by means of interlocking arrangements of certain units.

Branch circuit: That portion of a wiring system extending beyond the final overcurrent device protecting the circuit.

Brick ledge: A horizontal rectangular recess formed in a concrete wall to receive masonry.

Bridging: Lateral bracing of joists between supports to hold them in vertical alignment.

Built-up roofing: Roofing composed of two or more layers of roll roofing cemented together on the job. Sometimes called tar and gravel roofing.

Cantilever: When a structural element projects beyond its support it becomes a cantilever.

Carriage: The notched structural support for the steps of a flight of stairs.

Casement: An outswinging window with sash hinged on its sides like a door.

Centering: Temporary forms for the support of masonry arches or lintels.

Change Order: A revision of the original construction contract in writing signed by the architect and approved by the owner and the contractor.

Channel: A hot-rolled structural steel shape having a U-shaped section. Symbol C, or ⊏ now obsolete.

Chase: A channel on the inside of a wall for pipes, conduit, ducts, etc.

Column: A vertical supporting member loaded primarily in compression commonly made of wood, steel, or reinforced concrete.

Compression: A force that tends to shorten a member.

Contour line: A line on topographic maps representing points of constant height or elevation throughout its entire length.

Contractor: A person who does construction work under a contract.

Cornice: The ornamental projection of the roof at the top of a wall.

C. S. I.: The Construction Specification Institute.

Curtain Wall: An exterior wall that encloses a building but does not support floor or roof loads.

Details: Isolated and enlarged drawings used to explain the construction of a building.

Diazo Process: A dry reproduction process that makes a positive copy of a drawing with blue or black lines on a white background.

DO: Stands for ditto, the same repeated. Pronounced doe.

Dowel: A round pin of wood or steel used to hold or strengthen two pieces where they join. Wood dowels are used in furniture and millwork. Steel dowels are used at joints in concrete construction.

Eaves: The roof overhang or edge.

Elevation View: The side view of a building made using the orthographic projection method.

Engineer: A professionally qualified and duly licensed designer and administrator of civil, mechanical, electrical, chemical, or nuclear projects.

Face Brick: Bricks manufactured to meet special standards for color, texture, absorption, uniformity, and strength.

Fascia: The outermost flat vertical member of a cornice.

Fillet Weld: A triangular-shaped weld at the intersection of two surfaces at right angles to each other.

Flange: The projecting element parallel to the principle axis of structural shapes such as I-beams, channels, and tees.

Flat Plate Slab: Solid reinforced concrete floor system supported by columns without drop panels, capitals, or beams.

Flat Slab: Solid reinforced concrete floor system supported by columns with drop panels and capitals but no beams.

Fluorescent Lamp: An electric discharge type light source producing visible light by causing phosphorus to fluoresce when exposed to ultraviolet light produced by a mercury arc within the tubular bulb.

Footing: Enlarged concrete element designed to distribute column and foundation loads to the ground.

Format: The design, plan, or arrangement of anything.

Framing Plan: A plan view of a structure showing individual structural members in symbolic form.

Frieze: A decorative band at the top of a wall's surface, part of the cornice detail.

Gable: The triangular wall surface formed by two sloping roof surfaces at the end of a gable-roofed building.

Gambrel Roof: A roof with two pitches, each side of the ridge with the lower slope steeper than the upper.

Girder: The largest most important horizontal structural member carrying loads principally in bending.

Glu-lam: Heavy timber sections constructed by gluing many layers of lumber together. May be straight or curved.

Grade Beam: Reinforced concrete beams supported by piers to form the foundation for a building.

Grid: A pattern formed by sets of parallel lines running at right angles to each other. Used to organize construction drawings.

Grout: A fluid mixture of cement, sand, and water.

Heavy-Timber: A type of construction consisting of solid or laminated wood structural members of 4 x 6 minimum size with nominal 2 inch minimum decking. Floor and roof construction shall not have concealed spaces.

HID Lamps: High intensity discharge lamps including mercury, metal halide, and sodium light sources.

Hip Rafter: The diagonal rafter extending from the plate to the ridge at the intersection of two planes of a hip roof.

Hip Roof: A roof composed of sloping planes rising from all four sides of a building.

Hopper Vent: An inward opening sash hinged at its lower edge.

Hyperbolic Paraboloid: A thin-shelled geometric shape with a double-curvature surface generated by straight lines.

Incandescent Lamp: A light source producing light by heating a filament to incandescence when an electric current passes through it.

Isometric Drawings: Drawings made with the object turned and tilted so that all three faces appear equal.

Jamb: The exposed lining or side frame of a window or door opening.

Joist: Closely spaced horizontal structural members that carry loads in bending. They frame into beams or girders or onto walls.

Kiln Dried: Lumber dried by placing in a heated enclosure under carefully controlled conditions.

Landing: A platform between flights of stairs.

Lintel: A horizontal structural member used to support the wall above door and window openings.

Lookout: The horizontal framing members running from wall to roof edge in closed cornice construction.

Mansard Roof: A roof with a very steep slope, rising from the eave and then breaking to a low slope or flat plane. Forms hips at corners.

Mat Foundation: A foundation system in which the entire area of the building acts as a footing.

Modular: Relates to a system of coordination of the dimensions of buildings and the materials used to construct them through reference to a four-inch three dimensional unit.

Moulding: Ornamental strips used to enhance the appearance of building elements usually at the juncture of two surfaces.

Mullion: The trim dividing multiple windows.

Multi-view Drawings: Two or more related orthographic views of an object.

Nosing: The overhanging rounded edge of a stair tread.

o. c. or OC: On center.

Oblique Drawings: Three dimensional views with one face parallel to the picture plane and receding lines parallel to each other.

Orthographic Projection: A method for drawing an object with all lines of sight perpendicular to the picture plane.

Outlet: Connection points in an electrical system for power consuming or control devices.

Pedestal: A section of concrete at the base of a column used to spread the column load over a greater area of the footing.

Perspective: A drawing of an object made with lines of sight radiating from a point.

Piers: A solid support of masonry with a height not exceeding 10 times its least dimension. A drilled pier is a concrete filled, round hole in the ground used to support structures.

Pile: A long slender wood, concrete, or steel structural support driven into the ground.

Pitch: In carpentry, pitch is the ratio of the rise of a roof to its span expressed in the simplest fractional form, 1/4, 1/2, etc.

Plan: A horizontal section view of a building taken about four feet above the level shown.

Plates: Large flat rectangular pieces of steel more than eight inches in width.

Platform Framing: A system of framing wooden buildings using closely spaced 2 x 4 studs extending from one floor plane to the ceiling plane above. Each floor plane is a platform upon which the next wall is erected.

Plot Plan: A drawing showing the location of the construction in relation to the site.

Prestressed Concrete: Reinforced concrete in which the steel is stressed in tension before the member is put into service.

Primary Service: Electrical service at distribution voltages of 2400 V or above.

PSI: Pounds per square inch.

Receptacle: Wiring device that permits connection by means of an attachment plug.

Reglet: A groove in a wall to receive flashing.

Reinforcement: In concrete work, steel bars, fabric, or tendons cast in the concrete to enable it to carry greater loads.

Retaining Wall: A free-standing wall constructed to keep earth from sliding or falling.

Ribbon Board: A 1 x 6 board let-in the face of the studs to support joists in balloon-framed structures.

Ridge: The horizontal peak or highest part of a roof.

Riser: The vertical part of a stair step.

Rivet: A round steel fastener driven hot and secured by forming a head on the smooth end by riveting.

Sanitary Drainage System: The building drainage system that conveys the discharge from plumbing fixtures to the sewage disposal facility.

Scale Drawing: A drawing of an object usually smaller in size but in all parts proportional.

Scribe: To mark one part for cutting so as to fit neatly against another part, often of irregular profile.

Secondary Service: Electrical service at reduced voltages, usually 440 V or less.

Section: A drawing showing an object as if cut revealing the interior makeup.

Shapes: Hot-rolled structural steel sections such as angles, channels, I-beams, H-beams, tees, etc.

Shear Plate: Timber connectors used with bolts for joints between steel and wood parts or two wood parts to be assembled and taken apart several times.

Shear Wall: A wall incorporated into a structure to resist shear stresses in its own plane resulting from wind or earthquake forces.

Sheathing: The concealed layer of material fastened to the studs or rafters of a wood framed building.

Shop Drawing: Fabrication, erection, and setting drawings for materials and equipment manufactured away from the site. Catalog cuts, performance charts, brochures, and other data may be included. They are prepared by suppliers and manufacturers.

Sill: The horizontal beveled member at the bottom of a window or door frame. Also the first wood member placed on a foundation in wood framed buildings.

Soffit: The underside of building elements, except ceilings, usually part of the ornamental trim or covering of structural elements. The horizontal under surface of a closed cornice, also called a plancier.

Specs, specifications: The written instructions prepared primarily to describe the materials and workmanship of a construction project.

Split Ring Connector: A timber connector in the form of a hoop inserted in pre-cut grooves between two members and clamped with a bolt through the center of the connector.

Steel Grillage Foundation: A foundation for heavy steel column loads consisting of closely spaced steel beams running at right angles to each other.

Storm Drainage System: A drainage system that conveys rain water, cooling water, or similar liquid wastes to a point of disposal.

Stressed-skin Panels: Industrialized plywood construction in which sheathing is glued to wood joists forming a more efficient structural unit.

Tee: Structural steel shape made by splitting S, W, and M-shaped beams along the center line of their webs.

Tension: Stress in a structural member causing the member to stretch.

Terrazzo: A type of flooring composed of colored marble chips embedded in cement and ground to a smooth polished surface.

Thin-shell Structure: Domes, vaults, and hyperbolic shaped concrete construction using thin curved slabs.

Tilt-up Construction: Concrete wall construction in which the wall sections are cast horizontally at the site and hoisted into place.

Topographic Map: A map showing the surface features of land, especially contours and elevations.

Transformer: An electrical machine that converts alternating current of one voltage to another voltage higher or lower.

Transit: An instrument containing a telescope for sighting and scales for measuring horizontal and vertical angles.

Trap: A U-shaped device used in sanitary drainage systems to prevent the flow of sewer gas into the room while allowing the liquid waste to flow into the drainage piping.

Tread: The horizontal part of a stairstep.

Truss: A structural assembly of small parts in a system of related triangles to form a long spanning element; used to support roofs over clear spans of 30 to 300 feet.

Trussed Rafter: Light wood trusses spaced 24 inches on center used to clear span residential and small commercial buildings. Ceiling and roof materials are fastened directly to these units.

Waffle Slab: A reinforced concrete floor or roof with deep square voids on the underside to lighten the structure. Forms a two-way structure suitable for large spans.

Water Closet: A water flushing toilet.

Web: The connecting element between the flanges of structural steel shapes such as I-beams, channels, and tees. It is normal to the principle axis of the member.

Welded Wire Fabric: Cold drawn wire welded into an open rectangular or square pattern and used to reinforce concrete slabs. Formerly called welded wire mesh.

Wide-flange: An I-shaped steel section with extra heavy flanges now designated a W-shape.

Working Drawings: Drawings prepared for builders to use when constructing a building or other project.

Wythe: Each continuous vertical section of brick masonry.

BUILDING MATERIALS SYMBOLS

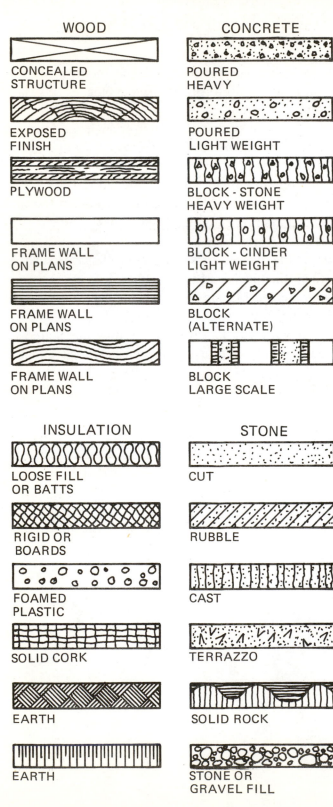

WOOD

CONCEALED STRUCTURE

EXPOSED FINISH

PLYWOOD

FRAME WALL ON PLANS

FRAME WALL ON PLANS

FRAME WALL ON PLANS

CONCRETE

POURED HEAVY

POURED LIGHT WEIGHT

BLOCK - STONE HEAVY WEIGHT

BLOCK - CINDER LIGHT WEIGHT

BLOCK (ALTERNATE)

BLOCK LARGE SCALE

BRICK

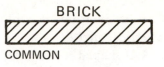

COMMON

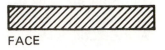

FACE

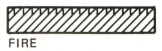

FIRE

STRUCTURAL CLAY TILE

TERRA COTTA

TILE ON CONCRETE

INSULATION

LOOSE FILL OR BATTS

RIGID OR BOARDS

FOAMED PLASTIC

SOLID CORK

EARTH

EARTH

STONE

CUT

RUBBLE

CAST

TERRAZZO

SOLID ROCK

STONE OR GRAVEL FILL

METAL, ETC.

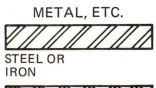

STEEL OR IRON

ALUMINUM

WELDED WIRE FABRIC & REBARS

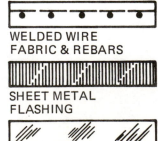

SHEET METAL FLASHING

GLASS IN ELEVATION

PLASTER, GROUT, STUCCO & MORTAR

Mechanical Symbols

Table of Contents

(By permission of the C. S. I., Inc., 1973, 1150 Seventeenth Street, N. W., Washington, D. C. 20036.)

Mechanical Reference Symbols

MISCELLANEOUS MECHANICAL ABBREVIATIONS

(Apply only when adjacent to a mechanical symbol or pipe size)

(May or may not be same as used in text of notes on drawings or in specifications.)

Air	A
Air Baseboard	AIR BB
Air Conditioning Unit	ACU
Air Handling Unit (AC or H&V)	AHU
Amps Full Load	FLA
Amps Locked Rotor	LRA
Architect/Engineer	A/E
Baseboard	BB
Cabinet Heater	CH
Cast Iron	CI
Ceiling Diffuser	CD
Cleanout	CO
Concrete	Conc
Convector	CONV
Difference In	$\triangle$
Dew Point Temperature	DPT
Entering Air Temperature	EAT
Exhaust Air	EA
Exhaust Fan	EF
Finned Tube Radiation	FTR
Fire Damper	FD
Hand-Off-Auto	HOA
Heating and Ventilating	H&V
Hose Bibb	HB
Invert	I
Latent Heat	LH
Latent Heat Ratio	LHR
Leaving Air Temperature	LAT

Mixed Air	MA
One Thousand British Thermal Units Per Hour	MBH
Outlet Velocity	OV
Outside Air	OA
Package Air Conditioning Unit	AC UNIT
Plumbing Fixtures	
Bath Tub	BT
Electric Water Cooler	EWC
Lavatory	L
Sink	S
Shower	SH
Urinal	U
Water Closet	WC
Pressure Drop	PD
Pump	P
Reheat Coil	RHC
Return Air	RA
Sensible Heat	SH
Static Pressure	SP
Supply Air	SA
Thermal Overload	TOL
Thermal Expansion Valve	TEV
Tip Speed	TS
Total Heat	TH
Undercut	UC
Unexcavated	UNEX
Unit Heater	UH
Unit Ventilator	UV
Utility Set	US
Velocity Pressure	VP
Vent Through Roof	VTR
Vitreous Clay Pipe	VCP

Wall Hydrant	WH
Wet Bulb Temperature	WB
Wide Flange	WF

MECHANICAL SYMBOLS

PIPES

Acetelyne	——AC——
Air Compressed	——A——
Air Compressed High Pressure	——HA——
Brine Return	———BR———
Brine Supply	——BS——
Chilled or Hot Water Return	———C/HR———
Chilled or Hot Water Supply	——C/HS——
Chilled Water Return	———CR———
Chilled Water Supply	——CS——
Cold Water	— — — —
Cold Water Lake	——L——
Cold Water Soft	——S——
Condensate Drain	——CD——
Condenser Water Return	———CWR———
Condenser Water Supply	——CWS——
Distilled Water	——DW——
Domestic Hot Water	— — — —
Domestic Hot Water (Special Water Temperature)	——180——
Domestic Hot Water Return	— — — —
Domestic Hot Water Return (Special Water Temperature)	——180——
Drain	——D——
Drinking Water Return	——DR——
Drinking Water Supply	——DS——
Fire Line	——F——
Fuel Oil Discharge	——FOD——
Fuel Oil Gauge	——FOG——
Fuel Oil Return	———FOR———
Fuel Oil Suction	——FOS——
Fuel Oil Tank Fill	——FOF——

Fuel Oil Tank Vent	———FOV———
Gas	——G——
Hot Water Heating Return	———HR———
Hot Water Heating Return High Temperature	———HHR———
Hot Water Heating Supply	——HS——
Hot Water Heating Supply High Temperature	——HHS——
Nitrous Oxide	——NO——
Oxygen	——O——
Pool Return	———PR———
Pool Supply	——PS——
Pump Discharge (Condensate or Vacuum Pump)	——PC——
Pump Discharge Feedwater	——PFW——
Pump Discharge to Boiler	——PBF——
Refrigerant Discharge	——HG——
Refrigerant Liquid	——RL——
Refrigerant Suction	———RS———
Sewer Combined *	——+——
Sewer Sanitary *	————————
Sewer Storm * *Sewers may be wider line outside building	— — —
Steam Condensate High Pressure Return	———HC———
Low Pressure Return	———C———
Medium Pressure Return	———MC———
Vacuum Return	———VC———
Steam High Pressure	——HS——
Steam Low Pressure	——S——
Steam Medium Pressure	——MS——
Vacuum	——V——
Vacuum Cleaning	——VC——
Vent	— — — —
Vent (Acid Resistant)	———AR———
Waste (Acid Resistant)	——AR——
Waste, Soil, or Leader	————————

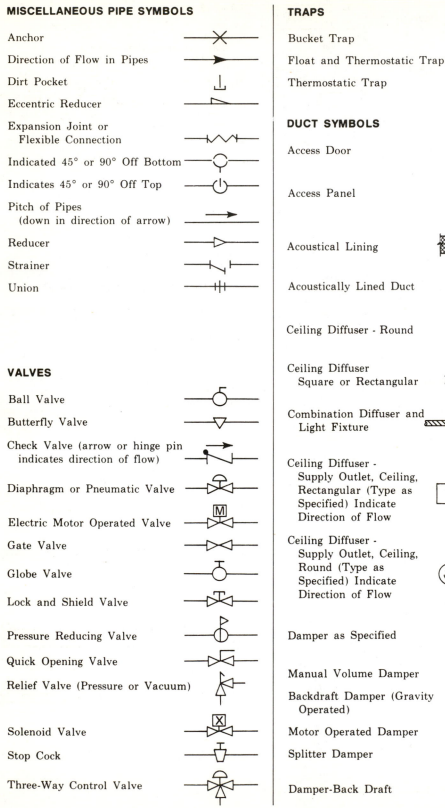

MISCELLANEOUS PIPE SYMBOLS

Anchor

Direction of Flow in Pipes

Dirt Pocket

Eccentric Reducer

Expansion Joint or
   Flexible Connection

Indicated 45° or 90° Off Bottom

Indicates 45° or 90° Off Top

Pitch of Pipes
   (down in direction of arrow)

Reducer

Strainer

Union

VALVES

Ball Valve

Butterfly Valve

Check Valve (arrow or hinge pin
   indicates direction of flow)

Diaphragm or Pneumatic Valve

Electric Motor Operated Valve

Gate Valve

Globe Valve

Lock and Shield Valve

Pressure Reducing Valve

Quick Opening Valve

Relief Valve (Pressure or Vacuum)

Solenoid Valve

Stop Cock

Three-Way Control Valve

TRAPS

Bucket Trap

Float and Thermostatic Trap

Thermostatic Trap

DUCT SYMBOLS

Access Door

Access Panel

Acoustical Lining

Acoustically Lined Duct

Ceiling Diffuser - Round

Ceiling Diffuser
   Square or Rectangular

Combination Diffuser and
   Light Fixture

Ceiling Diffuser -
   Supply Outlet, Ceiling,
   Rectangular (Type as
   Specified) Indicate
   Direction of Flow

12 x 12 CD
700 CFM
(CFM Optional)

Ceiling Diffuser -
   Supply Outlet, Ceiling,
   Round (Type as
   Specified) Indicate
   Direction of Flow

20φ CD
700 CFM
(CFM Optional)

Damper as Specified

Manual Volume Damper   VD

Backdraft Damper (Gravity
   Operated)   BDD

Motor Operated Damper   MOD

Splitter Damper   SD

Damper-Back Draft   BDD

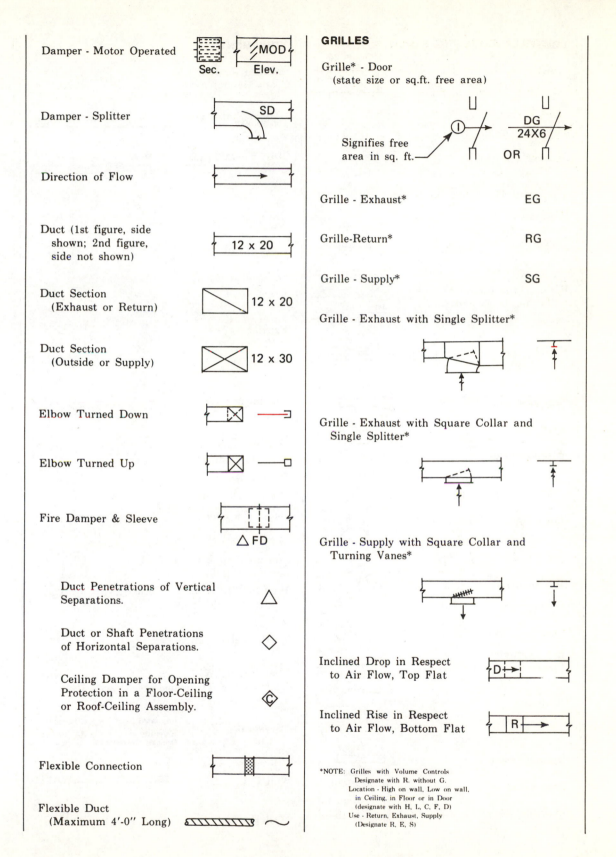

Damper - Motor Operated

Sec.   Elev.

Damper - Splitter

SD

Direction of Flow

Duct (1st figure, side
shown; 2nd figure,
side not shown)

12 x 20

Duct Section
(Exhaust or Return)

12 x 20

Duct Section
(Outside or Supply)

12 x 30

Elbow Turned Down

Elbow Turned Up

Fire Damper & Sleeve

△ FD

Duct Penetrations of Vertical
Separations.

Duct or Shaft Penetrations
of Horizontal Separations.

Ceiling Damper for Opening
Protection in a Floor-Ceiling
or Roof-Ceiling Assembly.

Flexible Connection

Flexible Duct
(Maximum 4'-0" Long)

GRILLES

Grille* - Door
(state size or sq.ft. free area)

Signifies free
area in sq. ft.

DG
24X6

OR

Grille - Exhaust*   EG

Grille-Return*   RG

Grille - Supply*   SG

Grille - Exhaust with Single Splitter*

Grille - Exhaust with Square Collar and
Single Splitter*

Grille - Supply with Square Collar and
Turning Vanes*

Inclined Drop in Respect
to Air Flow, Top Flat

D

Inclined Rise in Respect
to Air Flow, Bottom Flat

R

*NOTE: Grilles with Volume Controls
     Designate with R. without G.
     Location - High on wall, Low on wall,
        in Ceiling, in Floor or in Door
        (designate with H, L, C, F, D)
     Use - Return, Exhaust, Supply
        (Designate R, E, S)

LOUVERS

Louver (1st Figure, Side shown; 2nd Figure, side not shown)

36 × 24

Power or Gravity Exhaust

Power or Gravity Intake

Power or Gravity Roof Ventilator - Louvered

Radius Type Elbow (Splitter as specified)

Sound Trap   ST

Transformation

Turning Vanes

Unit Heater (Downblast)

Unit Heater (horizontal)

Unit Ventilator

(If recessed show amount)

Utility Set

Cabinet Heater

(If recessed show amount)

Air Handling Unit

EQUIPMENT

Air Vent or Relief Valve M=Manual A=Automatic   MAV AAV

Baseboard   WALL

Convector   SURFACE   SEMI-RECESSED   FULLY RECESSED

Drain, Area   AD

Drain, Floor   FD

Drain, Roof   RD

Existing

Fire Hydrant   FH

Humidistat   H

Outlet, Air   A

Outlet, Gas   G

Outlet, Vacuum   V

Pressure Tap & Gauge Cock   P

Pressure Tap & Gauge Cock w/O - 30 PSI Gauge   P 0-30

Thermometer Well   T

Thermometer Well w/0 - 100°F Thermometer   T 0-100

Thermostat   (Insulated Base) I   G (Guard) (Quantity of 4 T RA (Reverse Devices Controlled)   Acting)

Wall Fin   WALL

Wall Hydrant   WH

Electrical Symbols

Table of Contents

(By permission of the C. S. I., Inc., 1973, 1150 Seventeenth Street, N. W., Washington, D. C. 20036.)

Electrical Reference Symbols

ELECTRICAL ABBREVIATIONS

(Apply only when adjacent to an
electrical symbol)

Central Switch Panel	CSP
Dimmer Control Panel	DCP
Dust Tight	DT
Emergency Switch Panel	ESP
Empty	MT
Explosion Proof	EP
Grounded	G
Night Light	NL
Pull Chain	PC
Rain Tight	RT
Recessed	R
Transfer	XFER
Transformer	XFRMR
Vapor Tight	VT
Water Tight	WT
Weather Proof	WP

ELECTRICAL SYMBOLS

SWITCH OUTLETS

Single-Pole Switch	S
Double-Pole Switch	S_2
Three-Way Switch	S_3
Four-Way Switch	S_4
Key-Operated Switch	S_K
Switch and Fusestat Holder	S_FH
Switch and Pilot Lamp	S_P
Fan Switch	S_F
Switch for Low-Voltage Switching System	S_L
Master Switch for Low-Voltage Switching System	S_{LM}
Switch and Single Receptacle	⊖S
Switch and Duplex Receptacle	⊜S
Door Switch	S_D
Time Switch	S_T
Momentary Contact Switch	S_{MC}
Ceiling Pull Switch	Ⓢ
"Hand-Off-Auto" Control Switch	HOA
Multi-Speed Control Switch	M
Push Button	●

RECEPTACLE OUTLETS

Where weather proof, explosion proof, or other specific types of devices are to be required, use the upper-case subscript letters. For example, weather proof single or duplex receptacles would have the uppercase WP subscript letters noted alongside of the symbol. All outlets should be grounded.

Single Receptacle Outlet	⊖
Duplex Receptacle Outlet	⊜

Triplex Receptacle Outlet	⊕
Quadruplex Receptacle Outlet	⊕
Duplex Receptacle Outlet - Split Wired	⊖
Triplex Receptacle Outlet - Split Wired	⊕
250 Volt Receptable Single Phase Use Subscript Letter to Indicate Function (DW-Dishwasher; RA-Range, CD - Clothes Dryer) or numeral (with explanation in symbol schedule)	⊜
250 Volt Receptacle Three Phase	⊜
Clock Receptacle	Ⓒ
Fan Receptacle	Ⓕ
Floor Single Receptacle Outlet	⊖
Floor Duplex Receptacle Outlet	⊜
Floor Special-Purpose Outlet	⬙*
Floor Telephone Outlet - Public	◄
Floor Telephone Outlet - Private	◁

Example of the use of several floor outlet symbols to identify a 2, 3, or more gang floor outlet:

Underfloor Duct and Junction Box for Triple, Double or Single Duct System as indicated by the number of parallel lines.

Example of use of various symbols to identify location of different types of outlets or connections for underfloor duct or cellular floor systems:

Cellular Floor Header Duct

---

*Use numeral keyed to explanation in drawing list of symbols to indicate usage.

CIRCUITING

Wiring Exposed (not in conduit) ——E——

Wiring Concealed in Ceiling or Wall ————

Wiring Concealed in Floor — — — —

Wiring Existing* - - - - - - - -

Wiring Turned Up ————o

Wiring Turned Down ————●

Branch Circuit Home Run to Panel Board. → 2 1

Number of arrows indicates number of circuits. (A number at each arrow may be used to identify circuit number.)**

BUS DUCTS AND WIREWAYS

Trolley Duct*** | T | | T |

Busway (Service, Feeder, or Plug-in)*** | B | | B |

Cable Trough Ladder or Channel*** | C | | C |

Wireway*** | W | | W |

PANELBOARDS, SWITCHBOARDS AND RELATED EQUIPMENT

Flush Mounted Panelboard and Cabinet***

Surface Mounted Panelboard and Cabinet***

Switchboard, Power Control Center, Unit Substations (Should be drawn to scale)***

Flush Mounted Terminal Cabinet (In small scale drawings the TC may be indicated alongside the symbol)***

Surface Mounted Terminal Cabinet (In small scale drawings the TC may be indicated alongside the symbol)***

Pull Box (Identify in relation to Wiring System Section and Size)

Motor or Other Power Controller (May be a starter or contactor)***

Externally Operated Disconnection Switch***

Combination Controller and Disconnection Means***

POWER EQUIPMENT

Electric Motor (HP as indicated) ¼

Power Transformer

Pothead (Cable Termination)

Circuit Element, e.g., Circuit Breaker CB

Circuit Breaker

Fusible Element

Single-Throw Knife Switch

Double-Throw Knife Switch

Ground

Battery

Contactor C

Photoelectric Cell PE

Voltage Cycles, Phase Ex: 480/60/3

Relay R

Equipment Connection (as noted)

---

*Note: Use heavy-weight line to identify service and feeders. Indicate empty conduit by notation CO (conduit only).
**Note: Any circuit without further identification indicates two-wire circuit. For a greater number of wires, indicate with cross lines, e.g.:

—|—|—|— 3 wires;   —|—|—|— 4 wires, etc.

Neutral wire may be shown longer. Unless indicated otherwise, the wire size of the circuit is the minimum size required by the specification. Identify different functions of wiring system, e.g., signalling system by notation or other means.
***Identify by Notation or Schedule

REMOTE CONTROL STATIONS FOR MOTORS OR OTHER EQUIPMENT

Pushbutton Station	PB
Float Switch - Mechanical	F
Limit Switch - Mechanical	L
Pneumatic Switch - Mechanical	P
Electric Eye - Beam Source	
Electric Eye - Relay	
Temperature Control Relay Connection (3 Denotes Quantity.)	R₃
Solenoid Control Valve Connection	S
Pressure Switch Connection	P
Aquastat Connection	A
Vacuum Switch Connection	V
Gas Solenoid Valve Connection	G
Flow Switch Connection	F
Timer Connection	T
Limit Switch Connection	L

LIGHTING

	Ceiling	Wall
Surface or Pendant Incandescent Fixture PC = pull chain)	TYPE / WATTS	SWITCH / PC CIRCUIT
Surface or Pendant Exit Light		
Blanked Outlet	B	B
Junction Box	J	J
Recessed Incandescent Fixtures		
Surface or Pendant Individual Fluorescent Fixture		

Surface or Pendant Continous-
Row Fluorescent Fixture
(Letter indicating controlling switch)

Fixture No.
Wattage
Symbol not needed at each fixture

*Bare-Lamp Fluorescent Strip

ELECTRIC DISTRIBUTION OR LIGHTING SYSTEM, AERIAL

Pole**	
Street or Parking Lot Light and Bracket**	
Transformer**	
Primary Circuit**	
Secondary Circuit**	
Down Guy	
Head Guy	
Sidewalk Guy	
Service Weather Head**	

ELECTRIC DISTRIBUTION OR LIGHTING SYSTEM, UNDERGROUND

Manhole**	M
Handhole**	H
Transformer Manhole or Vault**	TM
Transformer Pad**	TP

Underground Direct Burial Cable
(Indicate type, size and number
of conductors by notation
or schedule)

Underground Duct Line
(Indicate type, size, and
number of ducts by cross-
section identification of each
run by notation or schedule.
Indicate type, size, and number
of conductors by notation or
schedule.

Street Light Standard Fed From
Underground Circuit**

---

*In the case of continuous-row bare-lamp fluorescent strip above an area-wide diffusing means, show each fixture run, using the standard symbol; indicate area of diffusing means and type by light shading and/or drawing notation.
**Identify by Notation or Schedule

SIGNALLING SYSTEM OUTLETS

INSTITUTIONAL, COMMERCIAL, AND INDUSTRIAL OCCUPANCIES

I   Nurse Call System Devices
     (any type)

Basic Symbol.

(Examples of Individual Item Identifiction Not a part of Standard)

Nurses' Annunciator
(add a number after it as
+①24 to indicate number
of lamps)

Call Station, single cord,
     pilot light

Call Station, double cord,
     microphone speaker

Corridor Dome Light, 1 lamp

Transformer

Any other item on same system -
     use numbers as required.

II   Paging System Devices
     (any type)

Basic Symbol.

(Examples of Individual Item Identification. Not a part of Standard)

Keyboard

Flush Annunciator

2-Face Annunciator

Any other item on same system -
     use numbers as required

III   Fire Alarm System Devices
     (any type) including Smoke and
     Sprinkler Alarm Devices

Basic Symbol.

(Examples of Individual Item Identification. Not a part of Standard)

Control Panel

Station

10″ Gong

Pre-signal Chime

Any other item on same system -
     use numbers as required.

IV   Staff Register System Devices
     (any type)

Basic Symbol.

(Examples of Individual Item Identification. Not a part of Standard)

Phone Operators' Register

Entrance Register - flush

Staff Room Register

Transformer

Any other item on same system -
     use numbers as required.

V   Electric Clock System Devices
     (any type)

Basic Symbol.

(Examples of Individual Item Identification. Not a part of Standard)

Master Clock

12″ Secondary - flush

12″ Double Dial - wall mounted

18″ Skeleton Dial

Any other item on same system -
     use numbers as required.

VI   Public Telephone System Devices

Basic Symbol.

(Examples of Individual Item Identification. Not a part of Standard)

Switchboard

Desk Phone

Any other item on same system -
     use numbers required.

VII Private Telephone System
     Devices (any type)

Basic Symbol.

(Examples of Individual Item Identification. Not a part of Standard)

Switchboard

Wall Phone

Any other item on same system -
use numbers as required.

VIII Watchman System Devices
      (any type)

Basic Symbol.

(Examples of Individual Item Identification. Not a part of Standard)

Central Station

Key Station

Any other item on same system -
use numbers as required.

IX Sound System

Basic Symbol.

(Examples of Individual Item Identification. Not a part of Standard)

Amplifier

Microphone

Interior Speaker

Exterior Speaker

Any other item on same system -
use numbers as required.

X Other Signal System Devices

Basic Symbol.

(Examples of Individual Item Identification. Not a part of Standard)

Buzzer

Bell

Pushbuttom

Annunciator

Any other item on same system
use numbers as required.

RESIDENTIAL OCCUPANCIES

Signalling system symbols for use in identifying
standardized residential-type signal system items
on residential drawings where a descriptive symbol list is not included on the drawing. When
other signal system items are to be identified, use
the above basic symbols for such items together
with a descriptive symbol list.

Pushbutton

Buzzer

Bell

Combination Bell-Buzzer

Chime

Annunciator

Electric Door Opener

Maid's Signal Plug

Interconnection Box

Bell-Ringing Transformer

Outside Telephone

Interconnecting Telephone

Television Outlet

Acknowledgments

I am grateful to those manufacturers and suppliers who furnished illustrations used in this material. The sources of such illustrations are listed below.

American Institute of Steel Construction, 25-7; 27-9

American Institute of Timber Construction, 18-2; 18-4; 18-12; 18-14

Andersen Corporation, 30-10

Brandom Kitchens, 29-6; 29-7

Bruning Division, Addressograph Multigraph, 1-4

Burris, L. Ryan (photographer), 1-1; 1-2; 1-3; 5-8; 5-9; 5-10; 16-13; 17-1

Concrete Reinforcing Steel Institute, 19-7; 19-8

Construction Specifications Institute, Appendix (Mechanical and Electrical references symbols)

Deer & Company, 11-1

Donn Products, Inc., 28-16

Johns-Manville, 28-17; 28-18

National Forest Products Association, 14-11; 14-12

Portland Cement Association, 22-11; 22-13

Stanpat Products, Inc., 6-14

Trus Joist Corporation, 15-9; 15-10

Wood connector Products, Inc., 16-8

I also thank my wife, Clara, who did all the manuscript typing.

Editorial

SOURCE EDITOR — Mark W. Huth

PROJECT EDITOR — Ann Mann